QUANT INTERVIEWS

FE PRESS

New York

Pocket Book Guides for Quant Interviews

1. 150 Most Frequently Asked Questions on Quant Interviews, Second Edition, by Dan Stefanica, Radoš Radoičić, and Tai-Ho Wang. FE Press, 2019

2. Probability and Stochastic Calculus Quant Interview Questions, by Ivan Matić, Radoš Radoičić, and Dan Stefanica. FE Press, 2021

3. Challenging Brainteasers for Interviews, by Radoš Radoičić, Ivan Matić, and Dan Stefanica. FE Press, 2023

Other Titles from FE Press

1. A Primer for the Mathematics of Financial Engineering, Second Edition, by Dan Stefanica. FE Press, 2011

2. Solutions Manual – A Primer for the Mathematics of Financial Engineering, Second Edition, by Dan Stefanica. FE Press, 2011

3. Numerical Linear Algebra Primer for Financial Engineering, by Dan Stefanica. FE Press, 2014

4. Solutions Manual – Numerical Linear Algebra Primer for Financial Engineering, by Dan Stefanica. FE Press, 2014

5. Elements of Stochastic Processes: A Computational Approach, by C. Douglas Howard. FE Press, 2017

6. A Probability Primer for Mathematical Finance, by Elena Kosygina.

CHALLENGING BRAINTEASERS FOR INTERVIEWS

RADOŠ RADOIČIĆ
IVAN MATIĆ
DAN STEFANICA

Baruch College
City University of New York

FE Press

New York

FE PRESS
New York

www.fepress.org

©Radoš Radoičić, Ivan Matić, Dan Stefanica 2023

All rights reserved. No part of this publication may be reproduced, stored in a retrieval system, or transmitted, in any form or by any means, electronic, mechanical, photocopying, recording, or otherwise, without the prior written permission of the publisher.

Printed in the United States of America

ISBN 978-1-7345312-1-3

To our beloved families

Contents

Preface ix

Acknowledgments xi

I Questions 1

1 Questions 3
- 1.1 "Aha" Questions 3
  - Quickies 3
  - Logical Conundrums 7
- 1.2 Probabilistic Puzzles 16
  - Discrete Probability 16
  - Continuous Probability 19
- 1.3 Combinatorial Puzzles 24
  - Counting Challenges 24
  - Games and Invariants 26
- 1.4 Tidbits from Other Areas of Mathematics . 30
  - Number Theory 30
  - Geometry 31
  - Calculus 32

vii

	Linear Algebra	34

II Solutions   37

2 "Aha" Questions   39
2.1 Quickies 39
2.2 Logical Conundrums 70

3 Probabilistic Puzzles   117
3.1 Discrete Probability 117
3.2 Continuous Probability. 165

4 Combinatorial Puzzles   215
4.1 Counting Challenges 215
4.2 Games, Invariants, Graphs, and Algorithmic Puzzles 232

5 Tidbits from Other Areas of Mathematics   259
5.1 Number Theory 259
5.2 Geometry 266
5.3 Calculus 277
5.4 Linear Algebra 313

6 Appendix   325
6.1 Sums of the form $\sum_{k=1}^{n} k^i$ 325

Bibliography   331

Preface

Brainteaser questions are an essential part of interviews across many firms and industries.

Checking the breadth and depth of domain knowledge is a matter that experts relish doing, and candidates understand its scope and how to prepare for it. On the other hand, brainteasers are meant to identify innovative thinking and the ability to make quick connections between seemingly separate knowledge areas.

This book is a resource to help candidates prepare for brainteaser questions on interviews for a large spectrum of industries, from quantitative finance, to technology and research, to science and engineering. A vast range of brainteasers is included in the book, from those with quick solutions that require a different way of looking at the problem, to logical conundrums, counting challenges, and mathematical brainteasers. We expect the reader to be intrigued and delighted. It is a feel-good book where the reader will learn serious matters while being entertained.

Interviewers should also find the book useful and interesting, finding a plethora of possible lines of questioning to ask candidates.

The brainteaser flavors included in the books are:

- "Aha Questions": Quickies and Logical Conundrums

- Probabilistic Puzzles: Discrete Probability and Continuous Probability

- Combinatorial Puzzles: Counting Challenges and Algorithmic Puzzles

- Mathematical Puzzles: Number Theory, Geometry, Calculus, and Linear Algebra

This is the third book in the *Pocket Book Guides for Quant Interviews* Series, following the second edition of the best–selling **150 Most Frequently Asked Questions on Quant Interviews**, and the probability–centered **Probability and Stochastic Calculus Quant Interview Questions**.

We hope you will enjoy and make the best use of our book.

Radoš Radoičić

Ivan Matić

Dan Stefanica

New York, 2023

Acknowledgments

As professors in the Financial Engineering Masters Program at Baruch College, it has been our privilege to have the opportunity to contribute to the career success of many cohorts of talented students and alumni. We take great pride in their achievements, which inspire us to find ever novel and creative ways to help them fulfill their tremendous potential. This book would have not been possible without our marvelous Baruch MFE community, and we are grateful to everyone who is part of it.

Working alongside our colleagues from the mathematics department and from the financial industry created the fertile environment that prompted us to write this book; we are thankful to all of them.

A special thank you is owed to the alumni and students who spearheaded the proofreading effort, and their help was greatly appreciated: Aneesh Subramanya, Bingqian (Aubree) Li, Chengxun (James) Wu, Gujia (Jane) Chu, Heqian (Adam) Wang, Jiahao Sun, Jialong (Jolene) Liao, Kai Zhang, Lei (Susan) Zhang, Ming Fu, Mingsen (Mason) Wang, Nicolas Buchwalder, Omar Twahir, Qiao Wang, Qichong Zhou, Shangwen Sun, Simon Trukhachev, Tianrong Wang, Weiye (Peter) Jiang, Xuan (Harper) Rui, Xiong Zhang, Xuebin (Alston) Gui, Yicheng Shi, Yizhou Wang, and Zhengjia Lu.

We are forever in debt to our families. (Special shout-out to Eric, as requested!) Their support and understanding made this book possible, and allowed you, the reader, to hold this book in your hands. Thank you for everything!

This book is dedicated to our families, from all our hearts.

Radoš Radoičić
Ivan Matić
Dan Stefanica

New York, 2023

xiii

Part I

Questions

Chapter 1

Questions

1.1 "Aha" Questions

Quickies

1. There are 100 numbers a_i, $i = 1, \ldots, 100$, such that each a_i is i less than the sum of the remaining 99 numbers. What is a_{50}?

2. You are given a rectangle with a smaller rectangle inside; note that the rectangles do not have to have parallel sides. Using one straight line, cut both of their areas into halves (simultaneously).

3. The time 12:21 is a palindrome, as it reads the same forward and backward. What is the shortest interval between two palindromic times?

4. Which number is larger, $\log_2(3)$ or $\log_3(5)$?

5. There are 15 bags on the table, each containing a different number of marbles. Determine the smallest possible number of marbles that could be on the table.

6. What is the smallest number of shots that has to be made to guarantee that a battleship (a 4×1 rectangle) hiding on a 10×10 board will be hit? The battleship can be located anywhere on the board and may be oriented either horizontally or vertically, and a shot is a blind guess of a square on the board. (Assume there are no other ships.)

7. A box contains 2021 red balls and 2021 blue balls. You remove two balls at a time repeatedly, and either discard both of them if they happen to be of the same color, or discard the blue ball and return the red ball if they happen to be of different colors. What is the probability that you will end up with one red ball in the box at the end?

8. (i) I give you an envelope that contains a certain amount of money, and you open it. I then put into a second envelope either twice this amount or half this amount, with a fifty–fifty chance of each. You are given the opportunity to trade envelopes. Should you?

   (ii) I put two sealed envelopes on a table. One contains twice as much money as the other. You pick an envelope and open it. You are then given the opportunity to trade envelopes. Should you?

9. Bob is a witty trader who trades exotic fruit grown far away. He travels from one place to another with three sacks which can hold 30 fruits each. None of the sacks can hold more than 30 fruits. He starts with 30 fruits in each sack. On his way, he must pass through 30 checkpoints, and at each checkpoint he has to give one fruit for each sack to the authorities. How many fruits remain after he goes through all the 30 checkpoints?

1.1. "AHA" QUESTIONS

10. Alice dives from a bridge into a river and swims upstream for one hour at a constant speed. She then turns around and swims downstream at the same speed. As Alice passes under the original bridge, a bystander tells her that her hat fell off into the river at the moment she dove in. In order to get her hat back, Alice continues to swim downstream without changing her speed. If she catches up to her hat when she is exactly one mile away from the bridge, what is the speed of the river?

11. Can you cover the entire two dimensional plane with the interiors of finitely many parabolas?

12. You are trapped on top of a building. The height of the building is 200 meters and you have a 150 meters rope. There is a hook at the top where you are standing. You know that there is a ledge with another hook midway between you and the ground. You have a sharp knife with you. How do you come down safely with the help of the rope, the knife, and the two hooks?

13. Two creeper plants, one red and the other green, are both climbing up and round a cylindrical tree trunk. The red plant twists clockwise and the green plant twists anti-clockwise. They both start at the same point on the ground. Before they reach the first branch of the tree, the red creeper made 5 complete twists, while the green creeper made 3 twists. Not counting the bottom and the top, how many times do the plants cross?

14. Alice shuffles her deck of cards, Bob shuffles his; then, Alice places her deck on top of Bob's. For each card in Alice's deck, count the number of cards between that card and its twin in Bob's deck. What is the sum of the numbers obtained?

15. The eight tiles numbered 1, 1, 2, 2, 3, 3, 4, 4 can be placed in a straight line such that there is one tile between the 1s, two tiles between the 2s, three tiles between the 3s, and four tiles between the 4s; for example, 4 1 3 1 2 4 3 2. However, it is impossible to do this if you add two 5. Can you explain why?

16. The n consecutive integers from 1 to n are written in a row. When can you place signs $+$ and $-$ in front of them so that the expression obtained is equal to 0?

17. You are given an 8×8 board filled with numbers 1 through 64, left to right, top to bottom, one number in each cell. Each number is then colored red or blue, so that every row and every column have exactly four red and four blue numbers. What is the difference between the sum of all the red numbers and the sum of all the blue numbers?

18. Ten thousand tiles are arranged in a square of the format 100×100. Each tile is colored red, blue, yellow or green, so that each 2×2 sub-square contains all four colors. What are the possible color combinations for the four tiles in the corners of the square?

19. Mr. and Mrs. Smith invited four other couples to a party. When everyone arrived, some people shook hands with some of the others. Nobody shook hands with his/her spouse or himself/herself and nobody shook hands with the same person twice. Afterwards, Mr. Smith asked everyone how many hands they shook and he received a different answer from each of them. How many hands did Mrs. Smith shake?

1.1. "AHA" QUESTIONS

20. Can you fill the entries of an 11×11 matrix with 0, +1, and −1, in such a way that the row sums and column sums are all different?

21. A cross-country runner runs a six miles course in 30 minutes. Show that somewhere along the course the runner ran a mile in exactly 5 minutes.

22. How many triangles of unit area with disjoint interiors can you fit inside a disk of unit radius?

23. Can you make an arithmetic progression with infinitely many terms such that all the terms are perfect squares?

24. Assume that
$$(x_i)_{i=1}^n \quad \text{and} \quad (y_i)_{i=1}^n$$
are two strictly decreasing sequences of real numbers. Find the permutation σ of $\{1, 2, \ldots, n\}$ that minimizes
$$\sum_{i=1}^n \left(x_i - y_{\sigma(i)}\right)^2.$$

25. Show that if all the points of the plane are colored red or blue, then there exists an equilateral triangle whose vertices are colored by the same color.

Logical Conundrums

1. A total of one hundred coins of various denominations lie in a row on a table. Alice and Bob alternately take a coin from either end of the row. Alice goes first which means that Bob will take the last coin on the table. Can Alice always guarantee to end up with at least as much money as Bob?

2. Is there a way to pack 250 bricks of the format $1 \times 1 \times 4$ into a $10 \times 10 \times 10$ box?

3. A coin rolls without slipping around another coin of the same size. How many times will the coin rotate (around its center) while making one revolution (around the other coin)?

4. After the revolution, each of the 66 citizens of a certain country, including the king, has a salary of $1. The king can no longer vote, but he does retain the power to suggest redistributions of salaries. Each salary must be a whole number of dollars, and the salaries must sum to $66. Each suggestion is voted on, and it is approved if there are more votes for than against. Each voter can be counted on to vote "yes" if his/her salary is to be increased, "no" if decreased, and otherwise not to bother voting. The king is both selfish and clever. What is the maximum salary he can obtain for himself, and how long does it take him to get it?

5. A kindergarten teacher has to arrange $2n$ children in n pairs for daily walks. Design an algorithm for this task so that no pair would be the same for $2n - 1$ days.

6. Given any sequence of 10 distinct integers, show that there exists either an increasing subsequence of length 4, or a decreasing subsequence of length 4.

7. Four sailors shipwrecked on a desert island gather coconuts into a large pile. They agree to share the coconuts equally and go to sleep for the nights. At different times during the night, each of them woke up, snuck to the coconut pile, counted it, determined it was one more than an exact multiple of 4, generously gave one coconut to a monkey, and buried

their share in a secret location on the island. If the pile never shrinks to fewer than 100 coconuts, what is the smallest possible number of coconuts that could have been in the original pile gathered by the sailors?

8. An equilateral triangle of side length 1 can be covered by 5 equilateral triangles of side length s. Can it also be covered by only 4 such triangles?

9. A 6×6 square is tiled by 2×1 dominoes that can be placed horizontally or vertically. Must there exist a line cutting the square without cutting any domino?

10. In this game, all the players start by sitting at a round table, each with a coin in their hand. At every turn, all the players simultaneously put their coin on the table in front of them. Each player can choose the side of the coin that will face down. Then, each player looks at the coins of his left and right neighbors. If the two coins show different faces of the coin, the player in the middle survives for the next turn. Otherwise, if both coins show the same face of the coin, the player in the middle lost.

After everybody knows who is continuing to the next turn, the losers pick up their coins and leave the table. The game is won when there is exactly one player left sitting at the table, who will be the winner and grab the lucky $10 bonus. If all the players are eliminated at the same time, the game is a tie and nobody gets the bonus. If there are only two players left, the players will necessarily eliminate each other at the next turn since left and right neighbors will be the same opponent.

You are offered a chance to play this game and grab the lucky $10 bonus. You can choose to play in a game starting with 10, 11, or 12 players, including yourself. What would you choose and why?

11. You are one of n people standing in a circle. Someone outside the circle goes around clockwise and repeatedly eliminates every other person in the circle, until one person – the winner – remains. Where should you stand to become the winner?

12. Four pebbles are initially placed on the ground so that they form a square. At each move, you can take an existing pebble from some point P and move it to a new point Q, as long as there is another pebble at the midpoint of PQ. Is it possible to form a larger square using the four pebbles after a finite number of moves?

13. An evil troll once captured a bunch of gnomes and told them: "Tomorrow, I will make you stand in a file, one behind the other, ordered by height such that the tallest gnome can see everybody in front of him. I will place either a white cap or a black cap on each head. Then, starting from the tallest, each gnome has to declare aloud what he thinks the color of his own cap is. In the end, those who were correct will be spared; the others will be eaten, silently."

The gnomes thought about it and came up with an optimal strategy. How many of them survived? What if the hats come in 10 different colors?

14. Two competitors play badminton. They play two games, each winning one of them. They then play a third game to determine the overall winner of the match. The winner of a game of badminton is the first player to score at least 21 points with a lead of at least 2 points over the other player. In this particular match, it is observed that the scores of each player listed in order of the games form an

1.1. "AHA" QUESTIONS

arithmetic progression with a nonzero common difference. What are the scores of the two players in the third game?

15. There are 25 people sitting around a table and each person has two cards. One of the numbers 1, 2, ..., 25 is written on each card, and each number appears on exactly two cards. At a signal, each person passes one of her cards, the one with the smaller number, to her right hand neighbor. Is it true that at some point one of the players will have two cards with the same numbers?

16. Alice can choose an arbitrary polynomial $p(x)$ of any degree with nonnegative integer coefficients. Bob can infer the coefficients of $p(x)$ by only two evaluations as follows: He chooses a real number a and Alice communicates $p(a)$ to him. He then chooses a real number b and Alice communicates $p(b)$ to him. What values of a and b could help *Bob* succeed and how?

17. I selected 3 positive integers: x, y, z. You can ask me about two linear combinations of these numbers with any coefficients: For example, you give me a, b and c and I tell you the value of $ax + by + cz$. What is your algorithm to find x, y and z?

18. An entry a_{ij} in a matrix is called a saddle point if it is strictly greater than all the entries in the i-th row of the matrix and strictly smaller than all entries in the j-th column of the matrix, or vice-versa. What is the maximum number of saddle points an $n \times n$ matrix can have?

19. 60 ambassadors are invited to a banquet. Every ambassador has at most 29 enemies. Can you seat all the ambassadors at a large round table such that nobody sits next to an enemy?

20. You are in a glitzy casino in Las Vegas. Having tried your hand at every game from Roulette to Black Jack, you managed to lose most of your money and have only one dollar left. What's worse, with all the champagne and everything, you misbehaved and the management made it very clear that you are not allowed to play any more games. However, you need two dollars to take the bus back to the hotel. Two shady characters at the bar offer you a game: they have a pile of 15 stones. Each of you in turn is to take your choice of 1, 2, 3, 4 or 5 stones from the pile. The person who takes the last stone gets one dollar from the person who drew previously, and the third person neither wins nor loses. You are to draw first. If both of the other players will play to their best personal advantage and will not make any mistakes, should you agree to play the game?

21. You have three identical Fabergé eggs. Each of them would break if dropped from the top of a 100–floors building. Your task is to determine the highest floor from which the eggs can be dropped without breaking. What is the minimum number of drops required to achieve this? You are allowed to break all the eggs in the process.

22. A rectangular table has 100 coins placed on it such that none of the coins overlap and it is impossible to put any more coins on the table without causing an overlap. Can you completely cover the table with 400 overlapping coins?

23. There are 100 coins on a table: 30 are genuine and 70 are fake. You know that all the genuine coins have the same weight; every fake coin is heavier than any genuine coin; and no two fake coins have the same weight. You have a balance with two pans (but no

1.1. "AHA" QUESTIONS

weights). What is the smallest possible number of weighings to identify at least one genuine coin?

24. Among 10 given coins, some may be real and some may be fake. All the real coins weigh the same. All the fake coins weigh the same, but have a different weight than the real coins. Can you prove or disprove that all ten coins weigh the same in three weighings on a balance scale?

25. A square grid with 10 rows and 10 columns contains the numbers 1, 2, ..., 100 in its squares written in that order from left to right and from bottom to top. The grid is then tiled with dominos. Each domino is given a value equal to the product of the numbers that are on it. What is the smallest possible sum of the values of the 50 dominos?

26. Four players sit in a circle on chairs numbered clockwise from 1 to 4. Each player has two hats, one black and one white, and is wearing one and holding the other. In the center sits a fifth player who is blindfolded. That player designates the chair numbers of those whose hats should be changed. His goal is to get all four wearing a hat of the same color, in which case the game stops. Otherwise, after each guess, the four walk clockwise past an arbitrary number of chairs (maintaining the same cyclic order), then sit for the next guess. Find a strategy that always works for the blindfolded player.

27. There are 100 prisoners in solitary cells. There is a central room with one light bulb that is initially off. No prisoners can see the light bulb from their cells. Every day, the warden picks a prisoner at random and that prisoner visits the central room. While there, the prisoner can toggle the bulb if he or she wishes. Also, the prisoner has the option of

asserting that all 100 prisoners have been to the central room by now. If this assertion is false, all 100 prisoners die. However, if it is indeed true, all the prisoners are set free. The prisoners are allowed to get together one night in the courtyard to discuss their strategy. What strategy should they agree on so that eventually someone can make a correct assertion? What is the expected time and variance of the time until they are all free?

28. There is a calculator in which all the digits (0, 1, ... 9) and the basic arithmetic operators (+, −, *, /) are disabled. However, other scientific functions are operational like *exp*, *log*, *sin*, *cos*, *arctan*, etc. The calculator currently displays a 0. Convert this first to 2 and then to 3.

29. Alice and Bob take turns picking up toothpicks off the floor. Alice goes first and she can pick up as many as she wants, but not all of them. In each subsequent turn, the person can pick up any number of toothpicks, as long as it does not exceed the number of toothpicks picked up in the previous turn. Passing your turn is not allowed. The person who picks up the last toothpick wins. Who will win?

30. A $p \times q \times r$ rectangular box is made from unit cubes. How many unit cubes does a diagonal of the cube go through?

31. You are presented with three large buckets. Each of them contains an integer number of ounces of a non–evaporating fluid. At any time you may double the contents of one bucket by pouring into it from a fuller one. Can you always empty one of the buckets eventually?

32. You die and awake in hell. Satan awaits you and has prepared a curious game. He arranged n quarters in

1.1. "AHA" QUESTIONS

a line, going in the east/west direction as follows: $THHH\ldots HHHT$. Satan explains the rules of the game: once a day, a coin is removed from the east end, and placed on the west end. If the coin was initially T, you get to choose whether the coin switches to H or not. If the coin was initially H, then Satan gets to make this choice. If at the end of one day all the coins are H, you get to leave Hell; Satan will of course try his hardest to make sure you never leave. Is there a strategy that allows you to eventually leave, or can Satan conspire to keep you in Hell forever?

33. If you hang a picture with string looped around two nails, and then remove one of the nails, the picture still hangs around the other nail. Can you come up with a different hanging of the picture with the property that removing *either* nail causes the picture to fall. Finally, can you hang a picture on three nails so that removing any one nail fells the picture?

34. A big rectangle is tiled by finitely many smaller rectangles, each having at least one side of integer length. Show that the big rectangle has at least one side of integer length.

1.2 Probabilistic Puzzles

Discrete Probability

1. There is an unknown total of m misprints in the book and the probability a reader spots a misprint, given that he is looking at it, is a constant. Two readers independently examine the page proofs of a book. Reader 1 flags 30 misprints, while Reader 2 flags 25 misprints. There are 5 flagged misprints in common. Find the number of remaining undetected misprints.

2. You are lost in a jungle, and you stop at a fork, wanting to know which of two roads lead to the village. Present are three willing natives, one each from a tribe of invariable truth-tellers, a tribe of invariable liars, and a tribe of random answerers. Of course, you don't know which native is from which tribe. You are permitted to ask only two yes-or-no questions, each question being directed to just one native. Can you get the information you need?

3. You and your opponent shall play a game with three dice: First, your opponent chooses one of the three dice. Next, you choose one of the remaining two dice. The player who throws the higher number with their chosen dice wins. Now, each dice has three distinct numbers between 1 and 9, with pairs of opposite faces being identical. Design the three dice such that you always win! In other words, no matter which dice your opponent chooses, one of the two remaining dice throws a number larger than your opponent, on average.

4. Bob chooses a number uniformly at random between 1 and 1000. Alice has to guess the chosen number as quickly as possible. Bob will let Alice know whether

1.2. DISCRETE PROBABILITY

her guess is smaller than, larger than or equal to the number. If Alice's guess is smaller than Bob's number, Bob replaces the number with another number chosen uniformly at random from $[1, 1000]$. Prove that there exists a strategy that Alice can use to finish the game in such a way that the expected number of steps is smaller than 45.

5. Given a regular 17-gon, choose 3 vertices at random and form a triangle. What is the probability that the center of the polygon is inside the triangle?

6. Find the probability that, in the process of repeatedly flipping a fair coin, you will encounter a run of 5 heads before a run of 2 tails.

7. Alice keeps tossing her fair coin until she gets 2 heads in a row. Bob keeps tossing his fair coin until he gets 3 heads in a row. Let A denote the number of Alice's tosses, and B denote the number of Bob's tosses. Find $\mathbb{P}(A > B)$.

8. Two players alternately roll an n-sided fair die. The player who fails to improve upon the previous roll loses. What is the probability that the first player wins?

9. There are 64 teams who play single elimination tournament, hence six rounds, and you have to predict the winners in all 63 games. Your score is then computed as follows: 32 points for correctly predicting the final winner, 16 points for each correct finalist, and so on, down to 1 point for every correctly predicted winner for the first round. (The maximum number of points you can get is thus 192.) Knowing nothing about any team, you flip fair coins to decide every one of your 63 bets. Compute the expected number of points.

10. Ten percent of the surface of a sphere is colored green and the rest is colored blue. Show that no matter how the colors are arranged, it is possible to inscribe a cube in the sphere so that all its vertices are blue.

11. A person has in each of his two pockets a box with n matches. Now and then he takes a match from a randomly chosen box until he finds the selected box empty. Find the expectation of the number of remaining matches in the other box.

12. The numbers $1, 2, \ldots, 2020$ on the x-axis are paired up at random to form 1010 intervals. Find the probability that one of these intervals intersects all the others.

13. You roll a fair 6-sided die and sum the numbers on the top face as long as you keep rolling larger and larger numbers on the top face. What is expected value of the sum once you stop rolling?

14. Roll six fair dice six times. Each time note the minimum. Find the expected value of the maximum amongst the six minima.

15. How many times do I have to roll a die on average until I roll the same number six times in a row?

16. A total of n balls, numbered 1 through n, are placed into n boxes, also numbered 1 through n in such a way that ball i is equally likely to go into any of the boxes $1, 2, \ldots, i$. Find the expected number of empty boxes.

17. What is the expected number of cards of a well-shuffled deck that would need to be turned until three spades appear?

1.2. CONTINUOUS PROBABILITY

18. At each stage, one can either pay $1 and receive a coupon that is equally likely to be any of the n types, or one can stop and receive a final reward of $3j$ if one's current collection of coupons contains exactly j distinct types. The objective is to maximize the expected net return. What is the optimal strategy?

19. Let's roll the die. If the outcome is 1, 2, or 3, we stop; otherwise, if it is 4, 5, or 6, a corresponding number of dice are rolled. For example, if the first roll gives 5, then in the second round we roll 5 dice, and so on. This procedure continues for every rolled dice whose outcome is 4, 5, or 6. What is the total expected sum of all the numbers obtained at the end of the Nth round of rolls?

20. We play the coin tossing game in which if tosses match, I get both coins; if they differ, you get both. You have m coins, I have n. What is the expected length of the game, that is, the number of tosses until one of us goes bankrupt?

Continuous Probability

1. Calculate
$$\int_{-\infty}^{+\infty} (2\pi)^{-\frac{1}{2}} e^{-\frac{t^2}{2}} \Phi(t+1)\, dt,$$
where $\Phi(\cdot)$ denotes the cumulative distribution function of a standard normal random variable.

2. Suppose that $X_1, X_2, \ldots, X_n$ are independent, identically distributed random variables. Show that
$$\frac{X_1^2}{X_1^2 + X_2^2 + \ldots + X_n^2}, \quad \frac{X_2^2}{X_1^2 + X_2^2 + \ldots + X_n^2}$$
are negatively correlated.

3. Let X, Y, Z be independent standard normal random variables. Find the distribution of the random variable
$$W = \frac{X + YZ}{\sqrt{1+Z^2}}.$$

4. Calculate
$$\int\int\cdots\int e^{-\sum_{1\leq i\leq j\leq n} x_i x_j}\, dx_1 dx_2 \ldots dx_n.$$

5. Let $A = (a_{i,j})$ and $B = (b_{i,j})$ be two symmetric non-negative semidefinite $n \times n$ matrices. Let $C = (c_{i,j})$ be the $n \times n$ matrix given by $c_{i,j} = a_{i,j} b_{i,j}$. Show that C is also symmetric non-negative semidefinite.

6. Find
$$\lim_{n\to\infty} \sum_{k=n}^{3n} \binom{k-1}{n-1} \left(\frac{1}{3}\right)^n \left(\frac{2}{3}\right)^{k-n}.$$

7. Let $X_1, X_2, \ldots, X_n$ be independent identically distributed uniform random variables on $[-1,1]$. Define the random variable
$$X = \frac{\sum_{i=1}^n X_i}{\sqrt{\sum_{i=1}^n X_i^2}}.$$
Approximate the probability of $X > 1$, as $n \to \infty$.

8. Calculate
$$\lim_{n\to\infty} \int_{[0,1]^n} \frac{x_1^5 + \ldots + x_n^5}{x_1^4 + \ldots + x_n^4}\, dx_1 \ldots dx_n.$$

9. Two real numbers X and Y are chosen at random in the interval $(0,1)$ with respect to the uniform distribution. What is the probability that the closest integer to X/Y is even? Express your answer in terms of π.

1.2. CONTINUOUS PROBABILITY

10. A random vector $X = (X_1, X_2, \ldots, X_n)$ is distributed uniformly at random on the unit sphere in $\mathbb{R}^n$. Find $\mathbb{E}\left[X_1^2\right]$ and $\mathbb{E}\left[X_1^4\right]$.

11. The number of people that have arrived at a train station by time t is a Poisson random variable with rate λt. If the train arrives at the station at a time uniformly distributed over $(0, T)$ (and independent of when the passengers arrive), what are the expected value and the variance of the number of passengers that enter this train?

12. An infinite sheet of paper has inscribed on it a set of horizontal lines D units apart and a set of vertical lines D units apart. A needle of length L (where $L < D$) is twirled and tossed on the paper. What is the expected number of lines crossed by the needle? What is the probability that the needle crosses a line?

13. Let $\{T_k\}$ be an independent identically distributed sample from the exponential distribution with parameter λ. Let $S_n = \sum_{k=1}^n T_k$ and define

$$N_t = \max\{k : S_k \leq t\}.$$

Find the distribution of N_t for $t > 0$.

14. Suppose that a building has a continuum of floors indexed by the real numbers in the interval $[0, 1]$. We think of 0 as the ground floor and 1 as the building's penthouse. The building has N elevators which at any given time are independently uniformly distributed along the building's floors. You are at floor a between 0 and 1, and push a button calling for an elevator. At that time all elevators head toward you at the same speed. What is the probability that the elevator that reaches you first is traveling downward?

15. Take a wire stretched between two posts and have a large number of birds land on it at random. Take a bucket of yellow paint and, for each bird, paint the interval from it to its closest neighbor. The question is: what proportion of the wire will be painted? More precisely: as the number of birds goes to infinity, what is the limit of the expected value of the proportion of painted wire, assuming a uniform probability distribution of birds on the wire?

16. In order to generate an $N(0,1)$-distributed random number, Bob proposes the following procedure: Using the $[0,1]$-uniform random number generator, independently generate X_1, X_2, ..., X_{12}, then compute $W = X_1 + X_2 + \ldots X_{12} - 6$. Bob claims W is approximately $N(0,1)$-distributed. Could you justify his claim? Compute $E[W^4]$. What do you think of his procedure?

17. Assume that there are two independently operated bus lines. Each of them stops in front of your apartment building. One stops every hour, on the hour. The other also stops once an hour, but not on the hour. That is, the first bus line arrives at ..., 6am, 7am, 8am, ...; while the second bus line arrives at ..., $6 + x$ am, $7 + x$ am, $8 + x$ am, ..., where x is a positive constant. Unfortunately, you don't know the value of x. Assume that x is uniform on $(0, 1)$. What is your average waiting time?

18. What is the probability that a point chosen at random from the interior of an equilateral triangle is closer to the center than to any of its sides?

19. In the segment $[0, 1]$, n points are chosen uniformly at random. For every point, one of the two directions (left or right) is chosen randomly and independently. At the same moment in time all n points

1.2. CONTINUOUS PROBABILITY

start moving in the chosen direction with speed 1. The collisions of all points are elastic. That means, after two points bump into each other, they start moving in the opposite directions with the same speed of 1. When a point reaches an end of the segment it sticks to it and stops moving. Find the expected time when the last point sticks to the end of the segment.

20. There are n points uniformly distributed in a unit disk. What is the median of the smallest distance between these points and the center of the disk?

1.3 Combinatorial Puzzles

Counting Challenges

1. Is it true that any set of ten distinct numbers between 1 and 100 contains two disjoint nonempty subsets with the same sum?

2. In how many ways, counting ties, can four horses cross the finishing line?

   (For example, two horses can finish in three ways: A wins, B wins, A and B tie.)

3. For a 6 × 4 rectangle, the square in the second row and third column is painted black. How many rectangles with sides along the grid lines contain the black square?

4. 30 senators attend a session. Each senator is an enemy to exactly 10 other senators. In how many ways can one form a 3-member committee so that either all of the senators on the committee are mutual enemies, or no two of the senators on the committee are enemies?

5. In a chess tournament, each two players played each other once. Each player got 1 point for a win, 1/2 point for a draw, and 0 points for a loss. Let S be the set of the 10 lowest-scoring players. If every player got exactly half his total score from the games played against the players from the set S, how many players were in the tournament?

6. The desks in a classroom with 39 desks are arranged in a rectangular 3 × 13 grid. In the morning, 39 students take seats, one per desk. After recess, the students take seats again, one per desk, in such a way that each student sits at a desk that is horizontally or vertically adjacent to the desk where they

1.3. COUNTING CHALLENGES

sat from before the recess. How many different seating arrangements are there after recess?

7. The desks in a classroom with 24 desks are arranged in a rectangular 2×12 grid. In the morning, 24 students take seats, one per desk. After recess, the students take seats again, one per desk, in such a way that each student sits at a desk that is horizontally or vertically adjacent to the desk where they sat from before the recess. How many different seating arrangements are there after recess?

8. How many 8×8 matrices with all entries 0 or 1 are there, such that the sum of each row and each column is odd?

9. A perfect in–shuffle of a deck of 52 cards is defined as follows: The deck is cut in half followed by interleaving of the two piles. So, if the cards were labeled 0, 1, 2, ..., 51, the new sequence is 0, 26, 1, 27, 2, 28, ... With repeated in–shuffles, do we ever get back the original order, and if so, after how many in–shuffles?

10. In how many ways can you divide 7 candies and 14 stickers among 4 children such that each child gets at least one candy and also gets more stickers than candies?

11. Let A and B be two sets with the property that there are exactly 144 sets that are subsets of at least one of A or B. How many elements are in the union of A and B?

12. Show that the number of ways of writing 2020 as a sum of distinct positive integers is equal to the number of ways of writing 2020 as a sum of positive odd integers.

13. Suppose that $n \geq 2$ and that $\{x_1, x_2, \ldots, x_n\}$ and $\{y_1, y_2, \ldots, y_n\}$ are two different sets. If the set of all pairwise sums $x_i + x_j$ is equal to the set of all pairwise sums $y_i + y_j$, i.e.,

$$\{x_i + x_j, 1 \leq i \neq j \leq n\}$$
$$= \{y_i + y_j, 1 \leq i \neq j \leq n\},$$

show that n is a power of 2.

Games, Invariants, Graphs, and Algorithmic Puzzles

1. A transformation of a polygon consists of cutting the polygon into two pieces by a straight line. Then, one of the pieces must be turned over and glued back to the other piece along the edge created by the cut.

   Is it possible to start with a square and after finitely many transformations end up with a triangle?

2. In the subtraction of one 3-digit number from another, Alice and Bob fill in the six digits in the following fashion: Alice chooses a number from 0 to 9 and Bob chooses where to enter it as a digit. They continue until all the blank digits are filled. Some numbers may appear more than once, leading zeroes are permitted as well. Alice is trying to maximize the difference, while Bob is trying to minimize it. If both players play their best, what is the difference?

3. Alice and Bob divide a pile of one hundred coins between themselves as follows: Alice chooses a handful of coins from the pile and Bob decides who will get them. This is repeated until all coins have been taken, or until one of them has taken nine handfuls. In the latter case, the other person takes all of

1.3. GAMES AND INVARIANTS

the remaining coins in the pile. What is the largest number of coins that Alice can be sure of getting?

4. There are four knights on the 3×3 chess board: the two white knights are at the bottom corners, and the two black knights are the two upper corners of the board. The goal is to switch the knights in the minimum number of moves so that the black knights are on the main diagonal and the white knights are on the other diagonal. How do you do it?

5. Two players start with the sequence 1, 2, ..., 101. They alternate the moves and in each of the moves a player chooses 9 of the numbers and removes them from the sequence. The game is over when only two numbers remain. The player that starts wins $x - 54$ dollars from the player that plays second; Here x is the difference between the remaining two numbers after 9 numbers were erased 11 times. Would Bob rather be the first or the second player?

6. In a computer game, a spy is located on a one-dimensional line. At time 0, the spy is at location a. With each time interval, the spy moves b units to the right if $b > 0$, and $|b|$ units to the left if $b < 0$. Both a and b are fixed integers, but they are unknown to you. Your goal is to identify the location of the spy by asking at each time interval (starting at time 0) whether the spy is currently at some location of your choosing. Devise an algorithm that will find the spy after a finite number of questions.

7. An infection spreads among the squares of an $n \times n$ checkerboard in the following manner: If a square has two or more infected neighbors, then it becomes infected itself. (For example, if one begins with all n infected squares on the diagonal, it spreads to the whole board eventually.) Can you infect the

whole board if you begin with fewer than n infected squares?

8. You have to make n pancakes using a skillet that can hold only two pancakes at a time. Each pancake has to be fried on both sides; frying one side of a pancake takes 1 minute, regardless of whether one or two pancakes are fried at the same time. Design an algorithm to do this job in the minimum amount of time. What is the minimum amount of time required?

9. Three spiders are trying to catch an ant. All are constrained to the edges of a cube. Each spider can move at least one third as fast as the ant can. Can the spiders always catch the ant?

10. You are given 8 unit cubes such that 24 of the faces are painted blue and 24 of the faces are painted red. Is it always possible to use these cubes to form a $2 \times 2 \times 2$ cube that has the same number of blue and red unit squares on its surface?

11. Jerry the mouse eats his way through a $3 \times 3 \times 3$ cube of cheese by eating all the $1 \times 1 \times 1$ sub–cubes. If Jerry starts at a corner sub–cube and always moves onto an adjacent sub-cube (sharing a face of area 1), can it do this and eat the center sub-cube last? Ignore gravity!

12. You want to invert a set of n upright cups by a series of moves in which $n-1$ cups are turned over at once. Show that this can always be done if n is even and it can never be done if n is odd.

13. Coins of various sizes are placed on a table, with some touching others. As many times as you wish, you may choose a coin and turn it over along with every other coin that it touches. If all the coins start

1.3. GAMES AND INVARIANTS

out showing heads, is it always possible to change them to all tails using these moves?

14. You have a checker in each of the following grid points: $(0,0)$, $(0,1)$, $(1,0)$, $(2,0)$, $(0,2)$, $(1,1)$. You can make a move as follows: if (x,y) is filled and $(x+1,y)$ and $(x,y+1)$ are both empty, then remove checker from (x,y) and put one checker at $(x+1,y)$ and one checker at $(x,y+1)$. Using these moves, can you remove the checkers from all of the six initial positions?

15. There are 100 rest stops in a forest and 1000 trails, each connecting a pair of rest stops. Each trail e has a level of difficulty $d(e)$; no two trails have the same difficulty. An intrepid hiker decided to spend a vacation by hiking 20 trails of ever increasing difficulty. Can he be sure that it can be done? He is free to choose the starting rest stop and the 20 trails form a sequence where the start of one trail is the end of a previous one.

16. A room has n computers, less than half of which are damaged. It is possible to query a computer about the status of any computer. A damaged computer could give wrong answers. How can you discover an undamaged computer?

17. There are $2n$ coins in a bag. Their values are 1, 2, ..., $2n$. Two players, A and B, alternate their moves. A starts the game. In each of the moves, the player takes a coin from the bag and puts in one of the two piles. After all coins are taken out, the player B takes the pile with more money. The player A takes the smaller pile. Let β be the amount of money that B receives. Let α be the money that A receives. What is $\beta - \alpha$ if both players play their best?

1.4 Tidbits from Other Areas of Mathematics

Number Theory

1. Find the smallest positive integer divisible by 225 whose only digits are 0 and 1.

2. You are given three piles with 5, 49, and 51 pebbles respectively. Two operations are allowed:

   (a) merge two piles together, or

   (b) divide a pile with an even number of pebbles into two equal piles.

   Is there a sequence of operations that would result in 105 piles with one pebble each?

3. Find the smallest positive value of $33^m - 7^n$, where m and n are positive integers.

4. Given any set of seventeen integers, show that there is at least one subset of nine integers whose sum is divisible by 9.

5. The integers greater than zero are painted such that every number is either red or blue. Both paints are used; blue number + red number = blue number, and blue number × red number = red number.

   Given only this information, for each of the following decide whether it is a blue number, a red number, or could be either:

   red number × red number?

   red number + red number?

   blue number × blue number?

   blue number + blue number?

1.4. TIDBITS FROM OTHER AREAS OF MATHEMATICS

6. Find all integer solutions of
$$x^3 + 2y^3 = 4z^3.$$

Geometry

1. You are given a finite number of points in the plane with the property that any line that contains two of these points contains at least three of these points. Show that all the points must lie on a straight line.

2. Find all functions $f : \mathbb{R}^2 \to \mathbb{R}$ defined over the two-dimensional plane with the property that, for any square $ABCD$,
$$f(A) + f(B) + f(C) + f(D) = 0. \qquad (1.1)$$

3. What is the largest possible value of n such that there exist n points in the plane with just two different distances between them?

4. The new campus of University College is a perfect disk of radius 1km. The Coffee Company plans to open 7 coffee shops on campus. Where do they have to be placed in order to minimize the maximum (straight-line) distance that a person anywhere on the campus has to walk to find a coffee shop?

5. Inscribe a regular N-gon in a circle of radius 1. Draw the $N-1$ segments connecting a given vertex to the $N-1$ other vertices. Find the product of the lengths of these segments.

6. Can you draw five rays from the origin in $\mathbb{R}^3$ so that every angle between two of the rays is obtuse?

Calculus

1. Given a polynomial f of degree 98 with the property that
   $$f(k) = \frac{1}{k}, \quad \forall\, k \in \{1, 2, \ldots, 99\},$$
   find $f(100)$.

2. How many real solutions does the following equation have:
   $$\sqrt[7]{x} - \sqrt[5]{x} = \sqrt[3]{x} - \sqrt{x}?$$

3. Find all the real solutions of the equation
   $$\sin(\cos x) = \cos(\sin x).$$

4. Let $f(x)$ be a positive, continuously differentiable function, defined for all real numbers, whose derivative is always negative. Fix any x_0 and run the Newton's method. What is $\lim_{n \to \infty} x_n$?

5. Find
   $$\lim_{n \to \infty} \left| \sin\left(\pi \sqrt{n^2 + n + 1}\right) \right|,$$
   where n is a positive integer.

6. Find the 100th derivative of the function
   $$\frac{x^2 + 1}{x^3 - x}.$$

7. Compute the integral
   $$\int_{-1}^{1} \frac{\sqrt[3]{x}}{\sqrt[3]{1-x} + \sqrt[3]{1+x}}\, dx.$$

8. Compute the integral
   $$\int (x^6 + x^3) \sqrt[3]{x^3 + 2}\, dx.$$

1.4. TIDBITS FROM OTHER AREAS OF MATHEMATICS

9. Compute the integral
$$\int_0^{\frac{\pi}{2}} \ln(\sin x)\, dx.$$

10. Compute the integral
$$\int_0^{\frac{\pi}{4}} \ln(1+\tan x)\, dx.$$

11. Compute the integral
$$\int_0^1 x^x\, dx.$$

12. Compute the integral
$$\int_0^\infty e^{-xy} \cdot \frac{\sin(ax)}{x}\, dx.$$

13. Compute the integral
$$\int_0^1 \int_0^1 \frac{1}{1-xy}\, dx\, dy.$$

14. Let $a > 0$, $b > 0$. Compute the integral
$$\int_0^\infty \frac{e^{-ax} - e^{-bx}}{x}\, dx.$$

15. Find the sum of the infinite series
$$\sum_{n=1}^\infty \frac{1}{2n^2 - n}.$$

16. Let k be a positive integer. Calculate
$$\sum_{n_1=1}^\infty \cdots \sum_{n_k=1}^\infty \frac{1}{n_1 \cdot \ldots \cdot n_k \cdot (n_1 + \ldots + n_k + 1)}.$$

17. Find a function that is equal to the sum of all of its derivatives.

18. Find the nonnegative solutions to the differential equation
$$yy'' + (y')^2 = yy'e^x.$$

19. When the clock strikes midnight, n bugs located at the vertices of a regular n-gon of unit side length begin crawling at equal speeds, in the clockwise directions, and directly towards the adjacent bug. They continue to do so until they all finally meet at the center of the polygon.

    What is the total distance traveled by each bug? How many times does each bug spiral around the center?

Linear Algebra

1. Let A and B be two square matrices of the same order that satisfy
$$A^{-1} + B^{-1} = (A+B)^{-1}.$$
Show that $\det(A) = \det(B)$.

2. Let M be a triangular $n \times n$ matrix with all the entries on or above the main diagonal equal to 1. Find a quick way to compute M^k, for $k \geq 1$.

3. The 300×300 matrix A has all the main diagonal entries equal to 17 and the rest of the entries are 11. What is $\det(A)$?

4. A bug starts at the vertex A of a triangle ABC. It then moves to one of its two adjacent vertices. How many paths of length 8 end back at vertex A? For example, one such path is $ABCABCABA$.

1.4. TIDBITS FROM OTHER AREAS OF MATHEMATICS

5. In determinant tic–tac–toe, Player 1 enters a 1 in an empty 3×3 matrix. Player 0 counters with a 0 in a vacant position, and the play continues in turn until the 3×3 matrix is completed with five 1s and four 0s. Player 0 wins if the determinant is 0, otherwise Player 1 wins. Assuming that both players pursue optimal strategies, who will win and how?

6. The city has a total of n citizens. The citizens are forming various clubs, and at some point the government decided to implement the following rules:

   1) Each club must have an odd number of members.

   2) Every two clubs must have an even number of members in common.

   Show that it is impossible to form more than n clubs.

Part II

Solutions

Chapter 2

"Aha" Questions

2.1 Quickies

Question 1. There are 100 numbers a_i, $i = 1, \ldots, 100$, such that each a_i is i less than the sum of the remaining 99 numbers. What is a_{50}?

Answer: Denote by

$$S = \sum_{i=1}^{100} a_i$$

the sum of the 100 numbers. Then, $a_i = (S - a_i) - i$, for all $i = 1, \ldots, 100$, and therefore

$$a_i = \frac{1}{2}(S - i), \quad \forall\ i = 1 : 100. \tag{2.1}$$

By summing up all the equations (2.1) over $i = 1, \ldots, 100$, and using the fact that $S = \sum_{i=1}^{100} a_i$, we obtain that

$$S = \frac{1}{2} \sum_{i=1}^{100} (S - i)$$

$$= 50S - \frac{1}{2} \sum_{i=1}^{100} i.$$

39

Thus,

$$49S = \frac{1}{2}\sum_{i=1}^{100} i = \frac{100 \cdot 101}{4} = 2525, \qquad (2.2)$$

where we used the fact that

$$\sum_{i=1}^{n} i = \frac{n(n+1)}{2}$$

for every positive integer n.

From (2.2), we obtain that $S = \frac{2525}{49}$, and therefore, from (2.1), we conclude that

$$a_{50} = \frac{1}{2}(S - 50) = \frac{75}{98}. \quad \square$$

Question 2. You are given a rectangle with a smaller rectangle inside; note that the rectangles do not have to have parallel sides. Using one straight line, cut both of their areas into halves (simultaneously).

Answer: Any line passing through the center of a rectangle (that is, through the intersection point of the diagonals of the rectangle) divides the rectangle into two congruent parts. Then, the line passing through the centers of the two rectangles will divide each rectangle into two equal halves. $\quad \square$

Question 3. The time 12:21 is a palindrome, as it reads the same forward and backward. What is the shortest interval between two palindromic times?

Answer: The times 9:59 and 10:01 are palindromic and two minutes apart. We will prove that they form the shortest interval between two palindromic times.

2.1. QUICKIES

Increasing a palindromic time by one minute cannot generate another palindromic time: If the last digit of the palindromic time is not 9 (e.g., 12:21), then one minute later only last digit of the time will change (e.g., 12:22) and the new time will not be a palindrome. Else, if the last digit of the palindromic time is 9 but the second to last digit is not 5 (e.g., 9:39), then one minute later the last digit will be 0 while the first digit will be 9 (9:40 for our example) and the new time will not be a palindrome. If the palindrome is 9 : 59, then one minute later is 10 : 00, which is not a palindrome. $\square$

Question 4. Which number is larger, $\log_2(3)$ or $\log_3(5)$?

Answer: Observe that

$$\begin{aligned}\log_2(3) &= \frac{1}{2}\log_2(9) > \frac{1}{2}\log_2(8) = \frac{3}{2}, \quad \text{and} \\ \log_3(5) &= \frac{1}{2}\log_3(25) < \frac{1}{2}\log_3(27) = \frac{3}{2}.\end{aligned}$$

Therefore, we have $\log_3(5) < \frac{3}{2} < \log_2(3)$. $\square$

Question 5. There are 15 bags on the table, each containing a different number of marbles. Determine the smallest possible number of marbles that could be on the table.

Answer: The minimum number of marbles on the table is 14. We will use nested bags: Put zero marbles in the first bag, then put the first bag inside the second bag and put one more marble in the second bag (but not inside the first bag), put the second bag inside the third bag and put one more marble in the third bag (but not inside the second bag), and so on until the 14–th marble is put inside the 15–th bag which will contain all the other 14 bags as well. $\square$

CHAPTER 2. "AHA" QUESTIONS

Question 6. What is the smallest number of shots that has to be made to guarantee that a battleship (a 4 × 1 rectangle) hiding on a 10 × 10 board will be hit? The battleship can be located anywhere on the board and may be oriented either horizontally or vertically, and a shot is a blind guess of a square on the board. (Assume there are no other ships.)

Answer: The smallest number of shots needed is 24.

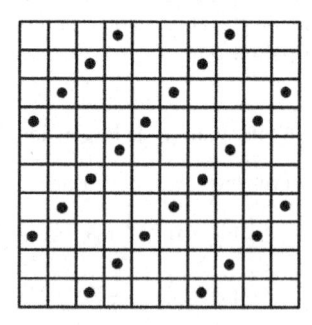

The figure above shows one possible placement of 24 shots that are to guaranteed to hit a battleship, regardless of where it is hiding on a 10 × 10 board. Indeed, every 4 × 1 rectangle, placed on a 10 × 10 board either horizontally or vertically, will contain one of the 24 black dots (shots).

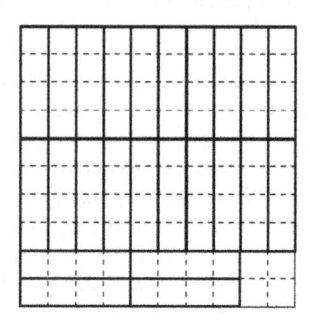

2.1. QUICKIES

Fewer than 24 shots will not suffice, since it is possible to place 24 battleships, with one shot being necessary to hit each of them, as shown in the figure above. □

Question 7. A box contains 2021 red balls and 2021 blue balls. You remove two balls at a time repeatedly, and either discard both of them if they happen to be of the same color, or discard the blue ball and return the red ball if they happen to be of different colors. What is the probability that you will end up with one red ball in the box at the end?

Answer: There is a 100% probability that you will end up with one red ball in the box at the end.

The only way you eliminate red balls is by removing two red balls at a time, if you picked two red balls from the box at the same time. Since you start with 2021 red balls, the number of red balls will always be odd and cannot become 0.

In the meanwhile, you will eliminate two blue balls every time you picked two blue balls, or you will eliminate one blue ball whenever you picked a red ball and a blue ball. This will happen until there are no blue balls left and all you will have left is an odd number of red balls. You will then continue to select and discard two red balls at a time until one red ball remains in the box. □

Question 8. (i) I give you an envelope containing a certain amount of money, and you open it. I then put into a second envelope either twice this amount or half this amount, with a fifty–fifty chance of each. You are given the opportunity to trade envelopes. Should you?

(ii) I put two sealed envelopes on a table. One contains twice as much money as the other. You pick an envelope and open it. You are then given the opportunity to trade envelopes. Should you?

Answer: (i) You should trade the envelopes. For simplicity, assume that the first envelope had $100 in it. Then, the other two envelopes hold $200 and $50, respectively. If you trade, the expected value of the amount you will have by choosing one of those two envelopes is

$$\frac{1}{2} \cdot \$200 + \frac{1}{2} \cdot \$50 = \$125,$$

which is higher than the $100 from the initial envelope.

(ii) In this case, you do not need to trade the envelopes. When you open the envelope you chose, you do not know whether it is the envelope with the higher amount or the envelope with the smaller amount. That information would become known only after trading envelopes, by which time it would be too late for the information to be useful. In other words, if, for example, the two envelopes held $100 and $200, the expected value of the amount you will have is $150 regardless of which envelope you would pick first. □

Question 9. Bob is a witty trader who trades exotic fruit grown far away. He travels from one place to another with three sacks which can hold 30 fruits each. None of the sacks can hold more than 30 fruits. He starts with 30 fruits in each sack. On his way, he must pass through 30 checkpoints, and at each checkpoint he has to give one fruit for each sack to the authorities. How many fruits remain after he goes through all the 30 checkpoints?

Answer: While Bob has fruits in all three sacks, he will have to give three fruits at each checkpoint. The optimal approach for Bob is to try to empty one bag as quickly as possible, so he will only have to give two fruits from that point on, then focus on emptying a second bag, after which he would only have to give one fruit at each checkpoint.

In other words, when passing through the first 10 checkpoints, Bob will give three fruits at each checkpoint from the same bag until that bag is empty. He is now left with two bags with 30 fruits each. When passing through the next 15 checkpoints, Bob will give two fruits at each checkpoint from the same bag until that bag is empty. He now has 30 fruits left, all of them in one bag. Bob has to go through 5 more checkpoints and will give one fruit at each of those checkpoints.

Bob will be left with 25 fruits, all of them in one bag. □

Question 10. Alice dives from a bridge into a river and swims upstream for one hour at a constant speed. She then turns around and swims downstream at the same speed. As Alice passes under the original bridge, a bystander tells her that her hat fell off into the river at the moment she dove in. In order to get her hat back, Alice continues to swim downstream without changing her speed. If she catches up to her hat when she is exactly one mile away from the bridge, what is the speed of the river?

Answer: First Solution: Denote by s_A Alice's swimming speed and by s_R the speed of the river flow. (Note that $s_A > s_R$ since Alice can swim upstream.)

The overall speed of Alice when she swims upstream (that is, her speed relative to the shore) is $s_A - s_R$ and therefore the distance swam upstream by Alice in $t_1 = 1$ hour is $D_1 = (s_A - s_R)t_1$. When Alice turns around and swims downstream, her overall speed is $s_A + s_R$. The time it takes Alice to come back to the original bridge when swimming downstream is

$$t_2 = \frac{D_1}{s_A + s_R} = t_1 \frac{s_A - s_R}{s_A + s_R}.$$

In the meantime, Alice's hat moved downstream at the river speed s_R for a time equal to $t_1 + t_2$. The distance between the hat and the bridge when Alice is back at the bridge is

$$\begin{aligned} \widetilde{D} &= s_R(t_1 + t_2) \\ &= s_R t_1 \left(1 + \frac{s_A - s_R}{s_A + s_R}\right) \\ &= \frac{2\, s_A\, s_R\, t_1}{s_A + s_R}. \end{aligned}$$

When Alice starts swimming downstream to recover her hat, her speed relative to the hat is $(s_A + s_R) - s_R = s_A$. The time it takes Alice to catch up to the hat is

$$t_3 = \frac{\widetilde{D}}{s_A},$$

and Alice will catch up to the hat at the distance

$$D_3 = t_3(s_A + s_R)$$

from the bridge, since her speed relative to the ground is $s_A + s_R$. Thus,

$$D_3 = t_3(s_A + s_R) = \frac{\widetilde{D}}{s_A}(s_A + s_R) = 2 s_R t_1. \quad (2.3)$$

We know that $D_3 = 1$ mile and $t_1 = 1$ hour. Then, from (2.3), it follows that $1 = 2 s_R$ and therefore $s_R = \frac{1}{2}$. We conclude that the speed of the river is 0.5 miles per hour.

Second Solution: Consider this problem from the perspective of the hat! Note that Alice's position relative to the hat's position is not affected by the speed of the current since both Alice and the hat are in the river. Alice swam away from the hat for one hour, so it takes another hour for her to swim back to the hat. In total, the hat has moved one mile from the bridge in two hours. Since the

2.1. QUICKIES

hat moved at the speed of the current and traveled one mile in two hours, we conclude that the speed of the current is $\frac{1}{2} = 0.5$ miles per hour.

Third Solution: Let us label by s_A and s_R the speeds of Alice and the river. We will calculate the total amount of time t_A that Alice spent in water before catching the hat. Then, we will calculate the total amount of time t_H that hat spent in water before beign caught by Alice. Finally, we will use $t_A = t_H$ to obtain the speed of the river.

Since the hat has traveled the distance 1 mile at the speed s_R, we obtain

$$t_H = \frac{1}{s_R}.$$

The time t_A can be written as $t_A = t_1 + t_2 + t_3$, where t_1 is the time that Alice swam upstream, t_2 is the time that Alice swam downstream before reaching the bridge, and t_3 is the time that Alice spent chasing the hat after meeting the bystander. The number t_1 is equal to 1 hour. We will write that as $t_1 = 1$. The distance that Alice traveled during that hour is $1 \cdot (s_A - s_R)$. Now, that distance will be covered at the speed $(s_A + s_R)$ which gives us

$$t_2 = \frac{s_A - s_R}{s_A + s_R}.$$

Finally, the number t_3 satisfies

$$t_3 = \frac{1}{s_A + s_R}.$$

The equation $t_1 + t_2 + t_3 = t_H$ becomes

$$\frac{1}{s_R} = 1 + \frac{s_A - s_R}{s_A + s_R} + \frac{1}{s_A + s_R}.$$

This equation is equivalent to $s_A = 2s_A s_R$. Since $s_A \neq 0$, we conclude that $s_R = \frac{1}{2}$, i.e. 0.5 miles per hour. □

Question 11. Can you cover the entire two dimensional plane with the interiors of finitely many parabolas?

Answer: We will use a proof by contradiction to show that it is not possible to cover the plane with interiors of finitely many parabolas.

We will first prove the following lemma:

Lemma. Assume that $m > 0$ and that $Q(a,b)$ is a point in the interior of the parabola $y = mx^2$. Every line through Q that is not parallel to the y-axis must intersect the parabola $y = mx^2$ at exactly two points.

Proof of the lemma. Let $\langle u, v \rangle$ be the vector of a line ℓ through $Q(a,b)$ that is not parallel to the y-axis. The parameter u must be non-zero. For every point (x, y) on the line ℓ, there exists a real number θ such that

$$\langle x, y \rangle = \langle a, b \rangle + \theta \langle u, v \rangle.$$

We will prove that there are always two values θ for which $(x(\theta), y(\theta))$ belongs to the parabola. In other words, we will prove that the equation $y(\theta) = mx^2(\theta)$ has two solutions in θ. The equation is equivalent to

$$\begin{aligned} b + \theta v &= m(a + \theta u)^2 \quad \Leftrightarrow \\ 0 &= mu^2 \theta^2 + (2mau - v)\theta + (ma^2 - b). \end{aligned}$$

We need to prove that the last quadratic equation in θ has a strictly positive discriminant. The discriminant satisfies

$$\begin{aligned} D &= 4m^2 a^2 u^2 - 4mauv + v^2 - 4mu^2(ma^2 - b) \\ &= 4mu^2 b - 4mauv + v^2. \end{aligned}$$

Since Q is in the interior of the parabola, we must have $b > ma^2$. Therefore,

$$D > 4m^2 u^2 a^2 - 4mauv + v^2 = (2mau - v)^2 \geq 0.$$

This completes the proof of the lemma.

2.1. QUICKIES

We will use the following obvious consequence of the lemma:

Consequence. If a ray (semi-line) ρ is fully in the interior of the parabola P, then ρ must be parallel to the axis of P.

Assume that there is a positive integer k and parabolas $P_1, \ldots, P_k$ whose interiors cover the entire plane. Observe that translation of a parabola along the line parallel to its axis results in a new parabola. One of the two parabolas is completely in the interior of the other one. Therefore, we may assume that each of the parabolas $P_1, \ldots, P_k$ contains the origin in its interior (those which don't would just get translated until they do).

For each $i \in \{1, 2, \ldots, k\}$, let us denote by ℓ_i the line that passes through the origin and that is parallel to the axis of the parabola P_i. Let us denote by ρ a line through the origin different from $\ell_1, \ldots, \ell_k$. The intersection of the line ρ with each of the parabolas is a bounded segment. The line ρ is unbounded, hence it must contain a point Z that is outside of the interiors of all parabolas. $\square$

Remark. Observe that a very similar argument can be used to show that the plane cannot be covered with the interiors of countably many parabolas.

Question 12. You are trapped on top of a building. The height of the building is 200 meters and you have a 150 meters rope. There is a hook at the top where you are standing. You know that there is a ledge with another hook midway between you and the ground. You have a sharp knife with you. How do you come down safely with the help of the rope, the knife, and the two hooks?

Answer: Let us label the top hook by T and the middle hook by M. The first step is to cut the rope into two pieces: One rope σ of length 50 and the other rope ρ of length 100. The idea is that once we descend to the ledge in the middle, we still remain in the full control of the rope ρ that has length 100.

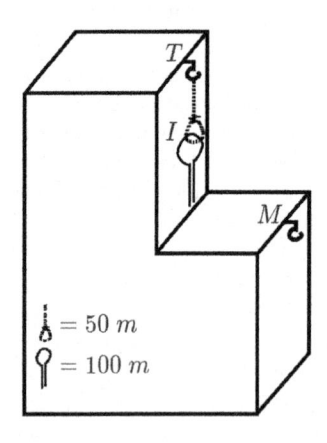

This would require us to wisely and skillfully use both ends of the rope σ of length 50. One end will be used to tie to the top hook T. The other end will be converted into an improvised hook on its own. We will use the letter I to denote this improvised hook. This way the rope σ of length 50 is actually providing us with the hook I that is at the distance 50 from the top of the building and at the height 150 from the ground.

We now have three hooks: One hook called M at the height 100 from the ground, the hook I at the height 150 from the ground, and the hook T at the height 200 from the ground. We can now come down safely with the rope ρ of length 100 as follows: First we put half of the rope ρ through the hook T and keep the ends of ρ together. This way the rope ρ is converted into a rope whose width is twice the original width and length is half of the original length. However, this is sufficient to descend to the hook

2.1. QUICKIES

I. Once we are at the hook I, then we can recover the rope ρ and repeat the procedure. Take half of the rope ρ through the hook I and use the obtained rope of length 50 to reach the ledge in the middle with the hook M. Once we reached the middle, we need to descend only once. The length 100 of the rope ρ is sufficient for this task. □

Question 13. Two creeper plants, one red and the other green, are both climbing up and round a cylindrical tree trunk. The red plant twists clockwise and the green plant twists anti-clockwise, both starting at the same point on the ground. Before they reach the first branch of the tree, the red creeper made 5 complete twists, while the green creeper made 3 twists. Not counting the bottom and the top, how many times do the plants cross?

Answer: The two creeper plants meet again after the red creeper made 5 complete twists and the green creeper made 3 twists. Imagine that we cut the trunk of the tree vertically from the common point where the two creeper plants begin from up to the point where the two plants meet again, and then unfold the bark. What we obtain is a rectangle with the vertical side of the rectangle being the unfolding of the circumference of the tree and the horizontal side of the rectangle corresponding to the vertical cut through the tree.

Assume, without loss of generality, that the red creeper starts at the lower left corner of the rectangle and the green creeper starts at the top left corner of the rectangle. For clarity, assume that the horizontal side of the rectangle has length 15 and the vertical side of the rectangle has length 1.

Note: in the picture below, the dashed line corresponds to the red creeper plant and the dotted line corresponds to the green creeper plant.

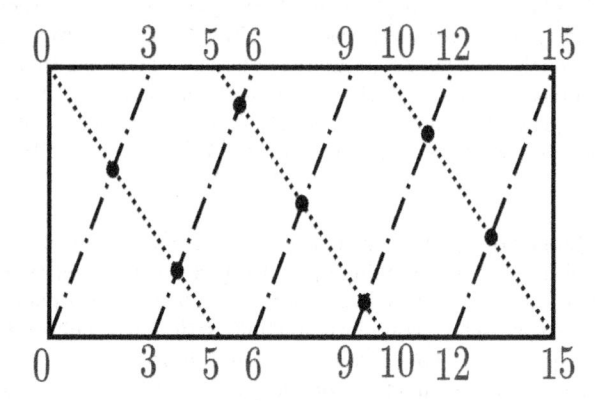

In the first twist, the red creeper goes from $(0,0)$ to $(3,1)$. The point $(3,1)$ on the upper side of the rectangle corresponds to the point $(3,0)$ on the lower side of the rectangle and they are exactly the same point on the tree since the upper horizontal line and the lower horizontal line both correspond to the vertical cut through the tree. Then, the second twist of the red creeper goes from $(3,0)$ to $(6,1)$, and the point $(6,1)$ on the upper side of the rectangle corresponds to the point $(6,0)$ on the lower side of the rectangle. Proceeding similarly, the other twists of the red creeper are from $(6,0)$ to $(9,1)$, from $(9,0)$ to $(12,1)$, and finally from $(12,0)$ to $(15,1)$.

Similarly, the green creeper starts at the top left corner $(0,1)$ of the rectangle. and its twists are as follows: from $(0,1)$ to $(5,0)$, then from $(5,1)$ to $(10,0)$, and finally from $(10,1)$ to $(15,0)$.

There are 7 intersection points of the lines going from the top side to the bottom side of the rectangle (the green creeper plant) and from the bottom side to the top side of the rectangle (the red creeper plant) and they correspond to the crossing points of the plants. □

2.1. QUICKIES

Question 14. Alice shuffles her deck of cards, Bob shuffles his; then, Alice places her deck on top of Bob's. For each card in Alice's deck, count the number of cards between that card and its twin in Bob's deck. What is the sum of the numbers obtained?

Answer: First Solution: Label the cards 1 through 52 in Bob's deck from bottom to top. For each $i \in \{1, 2, \ldots, 52\}$, denote by h_i the height of the card i in the Alice's deck. The heights of the cards in Alice's deck are 53, 54, ..., 104. Therefore, the sequence $(h_1, h_2, \ldots, h_{52})$ is a permutation of the sequence $(53, 54, \ldots, 104)$. The number of cards between the Bob's card i and the corresponding Alice's card is $(h_i - i - 1)$. Thus, the required sum of all the numbers is

$$\sum_{i=1}^{52}(h_i - i - 1) = -52 + \sum_{i=1}^{52} h_i - \sum_{i=1}^{52} i$$
$$= -52 + \sum_{i=1}^{52}(52 + i) - \sum_{i=1}^{52} i.$$

We can now combine the two sums from the last equation and obtain

$$\sum_{i=1}^{52}(h_i - i - 1) = -52 + \sum_{i=1}^{52}(52 + i - i)$$
$$= -52 + 52^2 = 2652.$$

Second Solution: Let us label the cards in the Bob's deck by $B_1, B_2, \ldots, B_{52}$. Here B_1 is the bottom card, and B_{52} is the top card. Label the cards in the Alice's deck by $A_1, \ldots, A_{52}$ in an analogous way. Denote by a_i the number of times that the card A_i is counted. Denote by b_i the number of times that the card B_i is counted. We need to determine

$$S = \sum_{i=1}^{52}(a_i + b_i).$$

Observe that A_i is counted exactly $52 - i$ times. It is counted every time we considered a card from the Alice's deck that was above A_i. Similarly, the card B_i is counted exactly $i - 1$ times. Hence we have $a_i = 52 - i$ and $b_i = i - 1$. The required sum S satisfies

$$S = \sum_{i=1}^{52}(52 - i + i - 1) = 51 \cdot 52 = 2652. \quad \square$$

Question 15. The eight tiles numbered 1, 1, 2, 2, 3, 3, 4, 4 can be placed in a straight line such that there is one tile between the 1s, two tiles between the 2s, three tiles between the 3s, and four tiles between the 4s; for example, 4 1 3 1 2 4 3 2. However, it is impossible to do this if you add two 5. Can you explain why?

Answer: Assume the contrary, that is, assume that it was possible to add two 5s. Label the positions of the 10 tiles by $1, 2, \ldots, 10$. Let a and $a + 2$ denote the positions of the 1s, b and $b + 3$ the positions of the 2s, c and $c + 4$ the positions of the 3s, d and $d + 5$ the positions of the 4s, and, finally, e and $e + 6$ the positions of the 5s. Then,

$$\begin{aligned}1 + 2 + \cdots + 10 &= a + (a + 2) + b + (b + 3) \\&\quad + c + (c + 4) + d + (d + 5) \\&\quad + e + (e + 6),\end{aligned}$$

yielding

$$55 = 2(a + b + c + d + e) + 20,$$

which is not possible since the left-hand side is odd, while the right-hand side is even. $\quad \square$

Question 16. The n consecutive integers from 1 to n are written in a row. When can you place signs $+$ and $-$ in

2.1. QUICKIES

front of them so that the expression obtained is equal to 0?

Answer: We will prove that we can choose the signs so that the expression obtained is equal to 0 if and only if n is divisible by 4 or $n + 1$ is divisible by 4.

First, we prove that the condition is necessary. Suppose that
$$\pm 1 \pm 2 \pm \ldots \pm n = 0,$$
for some choice of $+$ and $-$ signs. Moving the negative terms to the other side, we obtain a partition of the n consecutive integers from 1 to n into two subsets S and T such that $S \cup T = \{1, 2, \ldots, n\}$, $S \cap T = \emptyset$, and
$$\sum_{i \in S} i = \sum_{i \in T} i.$$
Since
$$\sum_{i \in S} i + \sum_{i \in T} i = \sum_{i=1}^{n} i = \frac{n(n+1)}{2},$$
we conclude that
$$\sum_{i \in S} i = \sum_{i \in T} i = \frac{n(n+1)}{4}.$$
Then, $n(n+1)$ must be divisible by 4 and therefore, either n is divisible by 4 or $n + 1$ is divisible by 4.

Next, we prove that the condition is sufficient. Let n be divisible by 4, that is, $n = 4k$ for some positive integer k. Then,
$$\sum_{j=0}^{k-1} ((4j + 1) - (4j + 2) - (4j + 3) + (4j + 4)) = 0.$$

Let $n + 1$ be divisible by 4, that is, $n = 4k - 1$, for some $k \in \mathbb{N}$. Then,
$$(1 + 2 - 3) + \sum_{j=1}^{k-1} (4j - (4j + 1) - (4j + 2) + (4j + 3)) = 0.$$

We conclude that the signs + and − can be placed in the described way if and only if n is divisible by 4 or $n+1$ is divisible by 4. □

Question 17. You have an 8×8 board filled with numbers 1 through 64, left to right, top to bottom, one number in each cell. Each number is then colored red or blue, so that every row and every column have exactly four red and four blue numbers. What is the difference between the sum of all the red numbers and the sum of all the blue numbers?

Answer: Label the rows 1 through 8, top to bottom, and label the columns 1 through 8, left to right. Let d denote the difference of the two sums. Since each row has exactly 4 red and 4 blue numbers, subtracting a fixed value from all of the numbers in a row does not change d. Hence, subtracting $(i-1) \cdot 8$ from all of the numbers in row i, for all $i = 1, 2, \ldots, 8$, does not change d.

Note that the column j of the board before the subtraction was

$$\begin{bmatrix} j \\ j+8 \\ j+16 \\ \vdots \\ j+48 \\ j+56 \end{bmatrix}.$$

The column j of the board after the subtraction is

$$\begin{bmatrix} j - 0 \cdot 8 \\ j+8 - 1 \cdot 8 \\ j+16 - 2 \cdot 8 \\ \vdots \\ j+48 - 6 \cdot 8 \\ j+56 - 7 \cdot 8 \end{bmatrix} = \begin{bmatrix} j \\ j \\ j \\ \vdots \\ j \\ j \end{bmatrix}.$$

2.1. QUICKIES

Thus, in the board obtained upon this subtraction, all of the numbers in column j are equal to j, for $j = 1, 2, \ldots, 8$. Since each column has exactly 4 red and 4 blue numbers, the difference between the sum of all the red numbers and the sum of all the blue numbers is equal to 0 in the board obtained upon this subtraction, and therefore it is equal to 0 in the initial board, since the subtraction did not change the difference. □

Question 18. Ten thousand tiles are arranged in a 100×100 square. Each tile is colored red, blue, yellow or green, so that each 2×2 sub-square contains all four colors. What are the possible color combinations for the four tiles in the corners of the square?

Answer: We will prove that all four corners must contain different tiles. Let us denote by r the number of corners that are red. It suffices to prove that $r = 1$. Let us first prove that

$$r \leq 2. \tag{2.4}$$

Assume that the tile at position $(1,1)$ is red. We will prove that tiles at positions $(100, 1)$ and $(1, 100)$ cannot be red. Let us prove that $(100, 1)$ is not red. The proof for $(1, 100)$ is the same. Without loss of generality, assume that a blue tile is at the position $(1, 2)$. We will now prove that $(100, 1)$ cannot be red nor blue. Since $(1, 1)$ and $(1, 2)$ are red and blue, the tiles in the second row that occupy the positions $(2, 1)$ and $(2, 2)$ must be green and yellow (in some order). We now conclude that the third row must have blue and red tiles at positions $(3, 1)$, and $(3, 2)$. Hence, the tiles at positions $(k, 1)$ and $(k, 2)$ are blue and red if and only if k is odd. Thus, $(100, 1)$ cannot be red. This completes the proof of (2.4). Let us assign the value 1 to each red tile and value 0 to every other tile. The sum of numbers in each 2×2 square is equal to 1. There are $99 \cdot 99$ squares of the format 2×2. If we add all the numbers in all these squares we obtain the sum $S = 99^2$. Observe that each corner is calculated only once when evaluating S. Each number on the edge that is not a corner is calculated exactly two times. Each number in the interior is calculated exactly 4 times. Thus, if we denote by c_1, c_2, c_3, and c_4 the numbers in four corners, then we have

$$S \equiv c_1 + c_2 + c_3 + c_4 \pmod{2}.$$

Since r is the total number of red squares in the corners, we have that $r = c_1 + c_2 + c_3 + c_4$. We can conclude that $S \equiv r \pmod{2}$. It remains to observe that $S = 99^2$ and $r \leq 2$. The only odd non-negative integer r that satisfies $r \leq 2$ must be equal to 1.

We have thus proved that exactly one corner is red. In an analogous way we prove that each color must appear exactly once in a corner. Obviously, each possible permutation of colors can be the colors in the corners. We partition the big square into disjoint 2×2 blocks and paint each block in the desired permutation. The corners will have the same pattern as the small 2×2 blocks. $\square$

Question 19. Mr. and Mrs. Smith invited four other couples to a party. When everyone arrived, some people shook hands with some of the others. Nobody shook hands with his/her spouse or himself/herself and nobody shook hands with the same person twice. Afterwards, Mr. Smith asked everyone how many hands they shook and he received a different answer from each of them. How many hands did Mrs. Smith shake?

Answer: We will prove that Mrs. Smith shook hands with 4 guests. There were 10 people in total, and no-one shook hands with the spouse. Therefore, the maximal number of

2.1. QUICKIES

handshakes that any person could have is 8. The number of handshakes for each person is an element of the set $\{0, 1, \ldots, 8\}$. This set has 9 elements. Since Mr. Smith received 9 different answers to his question, we conclude that except for Mr. Smith, the people had $0, 1, \ldots, 8$ handshakes. Let us denote by A_1 the guest who had 8 handshakes. The person A_1 shook hands with everybody except for the person who had 0 handshakes. Thus, the person with 0 handshakes must be the spouse of A_1. Let us denote this person by A_2.

We will now prove that the person with exactly 7 handshakes is the spouse of the person with exactly 1 handshake. Let us denote by B_1 the person with 7 handshakes and by B_2 the person with 1 handshake. The person B_2 only shook the hands with A_1. Thus, B_1 and B_2 did not shake hands. On the other hand, in order to have 7 handshakes, the person B_1 must have shaken hands with everyone except A_2 and B_2. Since A_2 is the spouse of A_1, it follows that B_2 must be the spouse of B_1. Using a similar reasoning we prove that the person C_1 with exactly 6 handshakes is the spouse of the person C_2 with exactly 2 handshakes and that the person C_1 shook hands with everyone except for A_2, B_2, and C_2. Continuing this line of reasoning leads us to conclude that the person D_1 with exactly 5 handshakes is the spouse of the person D_2 with exactly 3 handshakes.

We have now eliminated everyone and must conclude that the person with exactly 4 handshakes is Mrs. Smith. We can also conclude that Mr. and Mrs. Smith had exactly 4 handshakes each. Moreover, they shook hands with the same people: A_1, B_1, C_1, and D_1. □

Question 20. Can you fill the entries of an 11×11 matrix with 0, $+1$, and -1, in such a way that the row sums and column sums are all different?

Answer: Suppose that such a matrix, M, exists. Since the entries of M are all 0, $+1$, or -1, eleven row sums and eleven column sums are all different integers from $\{-11, -10, \ldots, 0, \ldots, 10, 11\}$, with exactly one of those integers, s, not appearing as a row or a column sum of M. We can assume $s \leq 0$, by symmetry. Then, there are exactly 11 rows and columns, in total, with a positive sum. Labeling the rows of M from top to bottom, and the columns of M from left to right, we can assume that the rows $1, 2, \ldots, r$ and the columns $1, 2, \ldots, 11 - r$ have positive sums, $0 \leq r \leq 11$. Let

- A be the sub-matrix formed by rows 1 through r and columns 1 through $11 - r$;

- B be the sub-matrix formed by rows 1 through r and columns $12 - r$ through 11;

- C be the sub-matrix formed by rows $r + 1$ through 11 and columns 1 through $11 - r$; and

- D be the sub-matrix formed by rows $r + 1$ through 11 and columns $12 - r$ through 11.

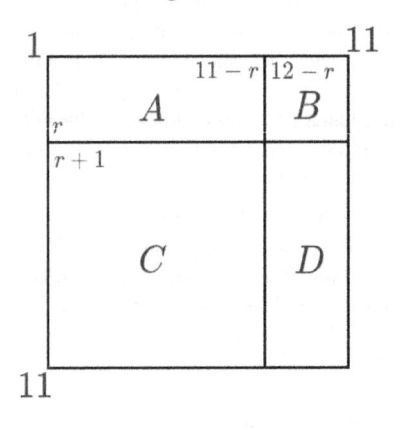

```
1                                    11
┌─────────────────────┬──────────────┐
│              11 − r │ 12 − r       │
│         A           │    B         │
│ r                   │              │
├─────────────────────┼──────────────┤
│ r + 1               │              │
│                     │              │
│         C           │    D         │
│                     │              │
│                     │              │
└─────────────────────┴──────────────┘
11
```

Denote by Σ_A, Σ_B, Σ_C, and Σ_D the sums of all the entries in matrices A, B, C, and D, respectively. Recall that the

2.1. QUICKIES

eleven row sums and eleven column sums span all the numbers from the set $\{-11, -10, \ldots, 0, \ldots, 10, 11\} \setminus \{s\}$. Thus, the sum of all the row entries of M plus the sum of all the column entries of M is

$$\left(\sum_{i=-11}^{11} i\right) - s = -s. \tag{2.5}$$

All the entries of M are summed up once in the sum $\Sigma_A + \Sigma_B + \Sigma_C + \Sigma_D$, and they are summed up twice in the sum of all the row entries and all the column entries of M. Therefore, from (2.5), we find that

$$\Sigma_A + \Sigma_B + \Sigma_C + \Sigma_D = -\frac{s}{2}. \tag{2.6}$$

Recall that $(\Sigma_A + \Sigma_B) + (\Sigma_A + \Sigma_C)$ is the sum of entries in all rows and columns of M that have a positive sum. Since the only missing sum is $s < 0$, it follows that

$$2\Sigma_A + \Sigma_B + \Sigma_C = \sum_{i=1}^{11} i = 66. \tag{2.7}$$

Then, from (2.6) and (2.7), we obtain that

$$\begin{aligned} -\frac{s}{2} &= \Sigma_A + \Sigma_B + \Sigma_C + \Sigma_D \\ &= (2\Sigma_A + \Sigma_B + \Sigma_C) - \Sigma_A + \Sigma_D \\ &= 66 - \Sigma_A + \Sigma_D. \end{aligned} \tag{2.8}$$

The matrix A has $r(11-r)$ entries, all of them at most 1. Hence,

$$\Sigma_A \leq r(11-r) \leq 30, \tag{2.9}$$

since the maximum of $r(11-r)$ with $r \in \{1, 2, \ldots, 11\}$ is obtained for $r = 5$ or $r = 6$ and is equal to 30. Similarly, the matrix D has $r(11-r)$ entries, all of them at least -1, and therefore

$$\Sigma_D \geq -r(11-r) \geq -30. \tag{2.10}$$

From (2.8) and using (2.9) and (2.10), we obtain that

$$-\frac{s}{2} \geq 66 - 30 - 30 = 6,$$

implying that $s \leq -12$, which is a contradiction.

We conclude that one cannot fill the entries of a 11×11 matrix with 0, $+1$, and -1, in such a way that the row sums and column sums are all different. $\square$

Question 21. A cross-country runner runs a six miles course in 30 minutes. Show that somewhere along the course the runner ran a mile in exactly 5 minutes.

Answer: Denote by $f(t)$ the distance in miles ran in t minutes by the runner. The function $f : [0, 30] \to [0, 6]$ is continuous and $f(0) = 0$, $f(30) = 6$. If the runner never ran a mile in exactly 5 minutes, then $f(t+5) - f(t)$ is never equal to 1 for any $0 \leq t \leq 25$. Since $f(t)$ is continuous, $f(t+5) - f(t)$ is also continuous. It follows that either $f(t+5) - f(t) > 1$ for all $0 \leq t \leq 25$, or $f(t+5) - f(t) < 1$ for all $0 \leq t \leq 25$. Both of these cases will lead to contradictions. If $f(t+5) - f(t) < 1$ for all $0 \leq t \leq 25$, then

$$f(30) < f(25)+1 < f(20)+2 < \cdots < f(5)+5 < f(0)+6,$$

which is not possible if $f(30) = 6$ and $f(0) = 0$.

If, $f(t+5) - f(t) > 1$ for all $0 \leq t \leq 25$, then

$$f(30) > f(25)+1 > f(20)+2 > \cdots > f(5)+5 > f(0)+6,$$

which, again, gives us the contradiction.

Thus, we conclude that the runner must have run a mile in exactly 5 minutes somewhere along the course. $\square$

2.1. QUICKIES

Question 22. How many triangles of unit area with disjoint interiors can you fit inside a disk of unit radius?

Answer: We can fit exactly two unit area triangles with disjoint interiors inside a disk of unit radius!

First, we will prove that the only unit area triangle that can fit inside a half-disk of unit radius is the isosceles right triangle with one of its sides being the diameter of the half-disk. Indeed, consider a triangle of maximum area that can fit inside a half-disk of unit radius. Since the area is maximal, the vertices of such a triangle must lie on the semicircle, and, furthermore, two of the vertices must be the endpoints of the diameter. Finally, among all triangles inscribed in a semicircle of unit radius, with one of the sides being the diameter, the isosceles right triangle has the maximum altitude, hence also the maximum area. The area of that triangle is exactly one.

Next, suppose that (at least) two unit area triangles with disjoint interiors are packed inside a disk of unit radius. According to the hyperplane separation theorem,[1] the two triangles can be separated by a line. This line cuts the disk into two regions, and one of them, containing one of the triangles is contained in a half-disk. According to the previous paragraph, this region must itself be a half-disk, and the triangle must be the isosceles right triangle with one of its sides being the diameter of this half-disk. Then, the other region must also be a half-disk, and the triangle must be the isosceles right triangle with one of its sides being the diameter of that half-disk. Note that we have not only proved that we can fit exactly two unit area triangles with disjoint interiors inside a disk of unit radius, but we have also shown that in every such packing of two unit area triangles with disjoint interiors inside a disk of

---

[1] The Hyperplane separation theorem is also called Minkowski's separation theorem. It states that any two convex disjoint sets in $\mathbb{R}^n$ can be separated by a hyperplane; see, e.g., [2] for more details.

64   CHAPTER 2. "AHA" QUESTIONS

unit radius, the two triangles must share the diameter of the disk, and their vertices taken together are the corners of a square inscribed in the circle of unit radius. □

Question 23. Can you make an infinite arithmetic progression with all the terms being perfect squares?

Answer: It is not possible to have an infinite arithmetic progression with all the terms being perfect squares.

We give a proof by contradiction. Assume that there exists an arithmetic progression with all the terms being perfect squares, and let d be the common difference between the two consecutive terms of the arithmetic progression.

We will show that the arithmetic progression cannot have any terms greater than d^2 and therefore must be finite. Else, if $x_k = a^2$ with $a > d$ is a term in the arithmetic progression, then $x_{k+1} \geq (a+1)^2$ and therefore

$$x_{k+1} - x_k \geq (a+1)^2 - a^2 = 2a + 1 > 2d + 1 > d.$$

Thus, x_k and x_{k+1} cannot be part of an arithmetic progression with common difference d. □

Question 24. Assume that

$$(x_i)_{i=1}^n \quad \text{and} \quad (y_i)_{i=1}^n$$

are two strictly decreasing sequences of real numbers.

Find the permutation σ of $\{1, 2, \ldots, n\}$ that minimizes

$$\sum_{i=1}^n \left(x_i - y_{\sigma(i)}\right)^2.$$

Answer: We will show by induction that the permutation that minimizes the sum

$$\sum_{i=1}^n \left(x_i - y_{\sigma(i)}\right)^2 \qquad (2.11)$$

2.1. QUICKIES

is $\sigma(i) = i$, for $i = 1 : n$. In other words, we will show that, if $\{x_i\}_{i=1:n}$ and $\{y_i\}_{i=1:n}$ are two strictly decreasing sequences, then the minimum of the sum (2.11) is obtained when x_i is paired with y_i, for all $i = 1 : n$, i.e.,

$$\min_{\sigma:\{1,2,\ldots,n\}\to\{1,2,\ldots,n\}} \sum_{i=1}^{n} \left(x_i - y_{\sigma(i)}\right)^2 \;=\; \sum_{i=1}^{n} (x_i - y_i)^2.$$

For $n = 2$, let $x_2 > x_1$ and $y_2 > y_1$. There are only two possible sums of the form (2.11) for $n = 2$. We will show that

$$(x_1 - y_1)^2 + (x_2 - y_2)^2 \;<\; (x_1 - y_2)^2 + (x_2 - y_1)^2, \quad (2.12)$$

and therefore conclude that the permutation that minimizes the sum (2.11) for $n = 2$ is $\sigma(1) = 1$ and $\sigma(2) = 2$, i.e.,

$$\min_{\sigma:\{1,2\}\to\{1,2\}} \sum_{i=1}^{2} \left(x_i - y_{\sigma(i)}\right)^2 \;=\; (x_1 - y_1)^2 + (x_2 - y_2)^2.$$

To do so, we will prove the following inequality which will be used again in the proof:

If $a > b$ and $c > d$, then

$$(a - c)^2 + (b - d)^2 \;<\; (a - d)^2 + (b - c)^2. \quad (2.13)$$

By expanding the squares in (2.13), we obtain

$$a^2 + b^2 + c^2 + d^2 - 2ac - 2bd \;<\; a^2 + b^2 + c^2 + d^2 - 2ad - 2bc,$$

which is equivalent to

$$ad + bc - ac - bd \;=\; (a-b)(d-c) \;<\; 0,$$

which is true, since $a > b$ and $c > d$.

For $a = x_1$, $b = x_2$, $c = y_1$, and $d = y_2$, it follows from (2.13) that

$$(x_1 - y_1)^2 + (x_2 - y_2)^2 < (x_1 - y_2)^2 + (x_2 - y_1)^2,$$

and therefore (2.12) is proved.

To complete the proof by induction, assume that, for any two strictly decreasing sequences of real numbers of length n, the permutation that minimizes the sum (2.11) is the identity permutation. Consider the strictly decreasing sequences $\{x_i\}_{i=1:n+1}$ and $\{y_i\}_{i=1:n+1}$ of length $n+1$, i.e., with

$$x_{n+1} < x_n < \cdots < x_2 < x_1; \quad y_{n+1} < y_n < \cdots < y_2 < y_1.$$

Let $\sigma : \{1, 2, \ldots, n, n+1\} \to \{1, 2, \ldots, n, n+1\}$ be the permutation that minimizes the sum

$$\sum_{i=1}^{n+1} \left(x_i - y_{\sigma(i)}\right)^2. \tag{2.14}$$

We first show that the permutation $\sigma : \{1, 2, \ldots, n, n+1\} \to \{1, 2, \ldots, n, n+1\}$ that minimizes (2.14) must have $\sigma(n+1) = n+1$. Else, let $1 \leq j \leq n$ such that $\sigma(n+1) = j$. Then, the minimum sum will contain the term $(x_{n+1} - y_j)^2$ and the other terms will come from pairing up the numbers from the following two strictly decreasing sequences of length n:

$$x_n < \cdots < x_{j+1} < x_j < x_{j-1} < \cdots < x_1;$$

$$y_{n+1} < \cdots < y_{j+1} < y_{j-1} < \cdots < y_1.$$

From the induction hypothesis, the permutation that minimizes the sum (2.14) must pair up x_n with y_{n+1}, x_{n-1} with y_n, and so on until x_j with y_{j+1}, and then x_{j-1} with y_{j-1}, x_{j-2} with y_{j-2}, and so on until x_1 with y_1.

2.1. QUICKIES

Thus, the minimum sum if $\sigma(n+1) = j$ (that is, if x_{n+1} is paired up with y_j) is

$$(x_{n+1} - y_j)^2 \tag{2.15}$$
$$+ \ (x_n - y_{n+1})^2 + (x_{n-1} - y_n)^2 + \cdots + (x_j - y_{j+1})^2$$
$$+ \ (x_{j-1} - y_{j-1})^2 + \cdots + (x_1 - y_1)^2. \tag{2.16}$$

Here, x_{n+1} is paired up with y_j and x_n is paired up with y_{n+1}. However, pairing up x_{n+1} with y_{n+1} and x_n with y_j would result in the smaller sum

$$(x_{n+1} - y_{n+1})^2 \tag{2.17}$$
$$+ \ (x_n - y_j)^2 + (x_{n-1} - y_n)^2 + \cdots + (x_j - y_{j+1})^2$$
$$+ \ (x_{j-1} - y_{j-1})^2 + \cdots + (x_1 - y_1)^2. \tag{2.18}$$

To see this, note that (2.17–2.18) is smaller than (2.15–2.16) if and only if

$$(x_{n+1} - y_{n+1})^2 + (x_n - y_j)^2 \ < \ (x_{n+1} - y_j)^2 + (x_n - y_{n+1})^2.$$

This follows from the inequality (2.13) with $a = x_n$, $b = x_{n+1}$, $c = y_j$, $d = y_{n+1}$, where $a > b$ and $c > d$ since $x_{n+1} > x_n$ and $y_{n+1} > y_j$.

We showed that any permutation with $\sigma(n+1) \neq n+1$, does not minimize the sum (2.14). Thus, the permutation

$$\sigma : \{1, 2, \ldots, n, n+1\} \to \{1, 2, \ldots, n, n+1\}$$

that minimizes (2.14) must have $\sigma(n+1) = n+1$.

Moreover, if $\sigma(n+1) = n+1$, we know from the induction hypothesis that the permutation of the set $\{1, 2, \ldots, n\}$ that minimizes the sum

$$\sum_{i=1}^{n} \left(x_i - y_{\sigma(i)}\right)^2$$

is exactly the identity permutation $\sigma(i) = i$ for all $i = 1 : n$. Therefore, the permutation

$$\sigma : \{1, 2, \ldots, n, n+1\} \to \{1, 2, \ldots, n, n+1\}$$

that minimizes (2.14) is also the identity permutation.

This completes the proof by induction of the fact that the given sum is minimized by choosing σ to be the identity permutation and therefore that

$$\min_{\sigma:\{1,\ldots,n\}\to\{1,\ldots,n\}} \sum_{i=1}^{n} \left(x_i - y_{\sigma(i)}\right)^2 = \sum_{i=1}^{n} (x_i - y_i)^2. \quad \square$$

Question 25. Show that if all the points of the plane are colored red or blue, then there exists an equilateral triangle whose vertices are colored by the same color.

Answer: We give a proof by contradiction. Assume that it is possible to color all the points of the plane red or blue such that no equilateral triangle would have the vertices colored with the same color.

Let $ABCDEF$ be a regular hexagon with center O and side length l. Without loss of generality, assume that the point O is colored red. The triangle $\triangle ACE$ is equilateral with side length $l\sqrt{3}$. Since we assumed that no equilateral triangle has vertices colored with the same color, at least one of the vertices of $\triangle ACE$ must be red. Without any loss of generality, assume that the vertex C is colored red. Then, the equilateral triangles $\triangle OBC$ and $\triangle OCD$ have two vertices, O and C, which are colored red, and therefore the vertices B and D must be blue if no equilateral triangle has vertices colored with the same color.

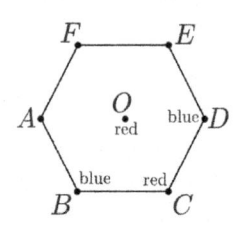

2.1. QUICKIES

Note that the triangle $\triangle BDF$ is equilateral with side length $l\sqrt{3}$. Since the vertices B and D are blue, the vertex F must be red. Then, the equilateral triangles $\triangle AOF$ and $\triangle EOF$ have two vertices, O and F, which are colored red, and therefore the vertices A and E must be blue.

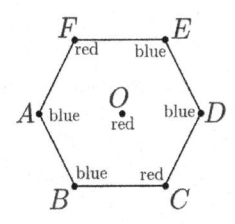

We now consider the regular hexagon $EOCGHJ$ with center D. The triangle $\triangle DEJ$ is equilateral and the vertices D and E are blue. This means that the vertex J must be colored red.

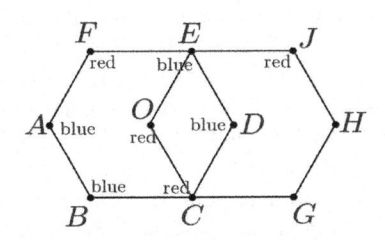

However, the triangle $\triangle CFJ$ is equilateral of side length 2 and all its vertices are colored red. This contradicts the assumption that it were possible to color all the points of the plane red or blue such that no equilateral triangle would have the vertices colored with the same color. □

2.2 Logical Conundrums

Question 1. A total of one hundred coins of various denominations lie in a row on a table. Alice and Bob alternately take a coin from either end of the row. Alice goes first which means that Bob will take the last coin on the table. Can Alice always guarantee to end up with at least as much money as Bob?

Answer: Yes! Label the positions of coins $1, 2, \ldots, 100$, from left to right, and let C_i denote the value of the coin in position i. Let S_E and S_O be the sums of values of all coins lying in even and odd positions. The numbers S_E and S_O are formally defined as

$$S_E = \sum_{i=1}^{50} C_{2i},$$

$$S_O = \sum_{i=1}^{50} C_{2i-1}.$$

The key observation is that, by taking the coin from an appropriate end of the row, Alice, being the player who makes the first move, can always make sure that Bob is never able to take the coin from an even position if $S_E > S_O$, thus winning the game with all the coins from even positions in hand.

Similarly, Alice can make sure that Bob is never able to take the coin from an odd position if $S_E < S_O$, thus winning the game with all the coins from odd positions in hand. If $S_E = S_O$, then, by taking the coin from an appropriate end of the row, Alice can play so as to either take all the coins in odd positions or all the coins in even positions, hence forcing a tie. $\square$

Question 2. Is there a way to pack 250 bricks of the format $1 \times 1 \times 4$ into a $10 \times 10 \times 10$ box?

2.2. LOGICAL CONUNDRUMS

Answer: No! Position the $10 \times 10 \times 10$ box in $\mathbb{R}^3$ so that the centers of its one thousand unit cubes have coordinates (i, j, k), with $1 \leq i, j, k \leq 10$. The set of all unit cubes with a fixed value of k, $1 \leq k \leq 10$, is said to form the level k of the box.

Color the unit cubes with four colors $\{0, 1, 2, 3\}$, so that the unit cube with center at (i, j, k) receives color $i+j+k$ (mod 4). Then, regardless of a direction in which a $1 \times 1 \times 4$ brick is placed into the $10 \times 10 \times 10$, it will contain exactly one unit cube of each of the four colors! In particular, if it were possible to pack 250 bricks of the format $1 \times 1 \times 4$ into a $10 \times 10 \times 10$ box, then the total number of unit cubes of each of the four colors would be exactly 250. However, it is easy to see that levels 1, 5, and 9 of the box each contain 26 unit cubes of color 0; levels 2, 4, 6, 8, and 10 of the box each contain 25 unit cubes of color 0; levels 3 and 7 of the box each contain 24 unit cubes of color 0, thus, making the total of 251 unit cubes of color 0. □

Question 3. If a coin rolls without slipping around another coin of the same size, how many times will it rotate (around its center) while making one revolution (around the other coin)?

Answer: The coin rotates exactly twice while making one revolution. Assume that each coin has radius 1. We will use the following fundamental principle of the car odometers:[2]

If the radius of the car tire is 1, then the tire makes one rotation if and only if its center (and, hence, the car) covers the distance 2π.

It doesn't matter whether the road is straight, curved, flat, or hilly. The only thing that can mess up the odome-

---
[2] A car odometer is the device that measures the distance traveled by the car which is shown on the dashboard.

ter is the ice. However, the problem was very explicit in clarifying that there is no slipping of the coins. Hence we can rely on the fact that when the car travels the distance 2π, the tire rotates exactly once.

We now focus our attention to the center of the revolving coin. The center travels around a circle of radius 2. Therefore, the center travels the distance $2 \cdot 2\pi = 4\pi$. Since the center of the coin covers the distance 4π, the coin rotates twice. □

Question 4. After the revolution, each of the 66 citizens of a certain country, including the king, has a salary of $1. The king can no longer vote, but he does retain the power to suggest redistributions of salaries. Each salary must be a whole number of dollars, and the salaries must sum to $66. Each suggestion is voted on, and it is approved if there are more votes for than against. Each voter can be counted on to vote "yes" if his/her salary is to be increased, "no" if decreased, and otherwise not to bother voting. The king is both selfish and clever. What is the maximum salary he can obtain for himself, and how long does it take him to get it?

Answer: The king can obtain $63 for himself, and it will take him 7 rounds of voting to achieve that.

First, the king arbitrarily selects 33 citizens and suggests they have their salaries increased to $2 each, at the expense of the remaining 33 citizens, including himself, who will have their salaries decreased to nothing. The 33 citizens, whose salaries are being doubled, are voting "yes", while the remaining 32 voting citizens (the king can no longer vote) are voting "no"; hence, king's suggestion of salary redistribution is approved. After the first round of voting, there are 33 citizens earning $2 each and 33 citizens (including the king) without a salary.

2.2. LOGICAL CONUNDRUMS

In the second round, the king suggests to increase of the salaries of 17 of the 33 salaried citizens to more than $2 each (so that their salaries sum to $66), while removing any salary from the remaining 16 (previously) salaried citizens. The 17 citizens, whose salaries are being increased, are voting "yes", the remaining 16 (previously) salaried citizens are voting "no", while the 32 voting citizens who lost their salaries in the first round do not even bother voting. Thus, again, king's suggestion of salary redistribution is approved. After the second round of voting, there are 17 citizens earning more than $2 each (for a total of $66) and 48 citizens (including the king) without a salary.

In the following four rounds, in the same manner, the king suggests reducing the number of salaried citizens to 9, 5, 3, and eventually 2, his suggestions being approved with the 9 to 8, 5 to 4, 3 to 2, and eventually 2 to 1 vote. After the sixth round of voting, there are 2 citizens earning all of the $66 salary in the country and 63 citizens (including the king) without a salary.

Finally, in the seventh round, the king selects 3 non-salaried citizens and suggests increasing their salaries to $1 each, while increasing his own salary to $63! The 3 previously non-salaried citizens are voting "yes", while the 2 citizens earning all of the $66 salary in the previous round are voting "no". The remaining citizens do not even bother voting, and the king's suggestion of salary redistribution is approved yet again. Therefore, after seven rounds of voting, the king finishes with the salary of $63 all to himself, which is clearly optimal! □

Question 5. A kindergarten teacher has to arrange $2n$ children in n pairs for daily walks. Design an algorithm for this task so that no pair would be the same for $2n - 1$ days.

Answer: Label the children from 1 to $2n$ arbitrarily. Two children in the same pair will be called "playmates." Place the child (labeled) 1 in the center of a circle. Place the children labeled 2 through $2n$ counterclockwise in $2n-1$ equidistant points on the circumference of the circle, that is, in the corners of a regular $(2n-1)$–gon inscribed in the circle. A set of pairs of children for a daily walk is obtained by drawing the diameter through the center and one of the points on the circumference, yielding one pair of children, while the other $n-1$ pairs are formed by the chords perpendicular to this diameter. Pairings for walks on other days are obtained by rotating the diameter, thus giving a new playmate for the center (child 1), and a new set of chords perpendicular to the new diameter, hence yielding new playmates for other children as well.

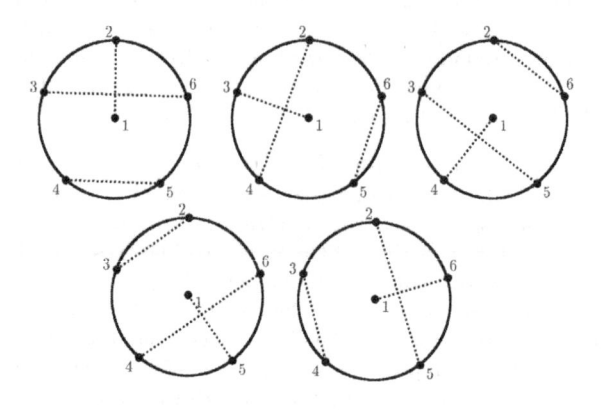

The case $n = 3$ is worked out in detail in the figure above.
□

Question 6. Given any sequence of 10 distinct integers, show that there exists either an increasing subsequence of length 4, or a decreasing subsequence of length 4.

2.2. LOGICAL CONUNDRUMS

Answer: We will prove this by contradiction. Suppose there is a sequence of 10 distinct integers, $\{a_i\}$, $i = 1, \ldots, 10$, that has no increasing subsequence of length 4 and no decreasing subsequence of length 4.

For $i = 1 : 10$, label a_i with the pair (m_i, n_i), where m_i is the length of the longest increasing subsequence ending with a_i, while n_i is the length of the longest decreasing subsequence ending with a_i.

Note that any two numbers in the sequence receive different labels. Indeed, for each pair (i, j) of different indices such that $i < j$, either $a_i < a_j$, or $a_i > a_j$. If $a_i < a_j$, then $m_i < m_j$. If $a_i > a_j$, then $n_i < n_j$.

Since $1 \leq m_i \leq 3$ and $1 \leq n_i \leq 3$ for all $i = 1, \ldots, 10$, there are only 9 possible labels, so by the pigeonhole principle, there exist two numbers in the sequence that received the same label, which is a contradiction.

This problem is a special case of a more general result, known as the Erdős–Szekeres theorem, which states that, given positive integers r and s, every sequence of $(r-1)(s-1)+1$ distinct real numbers contains an increasing subsequence of length r or a decreasing sequence of length s. The proof is essentially the same to the one given above. □

Question 7. Four sailors shipwrecked on a desert island gather coconuts into a large pile. They agree to share the coconuts equally and go to sleep for the nights. At different times during the night, each of them woke up, snuck to the coconut pile, counted it, determined it was one more than an exact multiple of 4, generously gave one coconut to a monkey, and buried their share in a secret location on the island. If the pile never shrinks to fewer than 100 coconuts, what is the smallest possible number of coconuts that could have been in the original pile gathered by the sailors?

Answer: Letting n denote the total number of the coconuts in the original pile, the sailor first to wake up took away a coconuts, where $n = 4a + 1$, and left a pile with $3a$ coconuts. Denoting by successive letters the shares taken by other sailors, we find that

$$n = 4a + 1; \quad 3a = 4b + 1 \qquad (2.19)$$
$$3b = 4c + 1; \quad 3c = 4d + 1, \qquad (2.20)$$

where n, a, b, c, and d are positive integers. After the fourth sailor buried his share, the pile with $3d$ coconuts is the smallest pile obtained in the process and must contain at least 100 coconuts. Therefore, $3d \geq 100$. The equations (2.19) and (2.20) can be rewritten as

$$n + 3 = 4(a + 1); \quad 3(a + 1) = 4(b + 1) \qquad (2.21)$$
$$3(b + 1) = 4(c + 1); \quad 3(c + 1) = 4(d + 1). \qquad (2.22)$$

From (2.21) and (2.22), it follows that

$$\begin{aligned} n + 3 &= 4(a+1) = 4 \cdot \frac{4}{3}(b+1) \\ &= 4 \cdot \left(\frac{4}{3}\right)^2 (c+1) \\ &= 4 \cdot \left(\frac{4}{3}\right)^3 (d+1). \end{aligned} \qquad (2.23)$$

Since $n+3$ is an integer, $d+1$ must be divisible by $3^3 = 27$. The smallest such value of d that satisfies $3d \geq 100$ is $d = 53$. Then, from (2.23), it follows that $n = 509$.

We conclude that the smallest possible number of coconuts that could have been in the original pile gathered by the sailors is 509. □

Question 8. An equilateral triangle of side length 1 can be covered by 5 equilateral triangles of side length s. Can it also be covered by only 4 such triangles?

2.2. LOGICAL CONUNDRUMS

Answer: Let A, B, and C be the vertices of the equilateral triangle of side length 1 and denote by D, E, and F the midpoints of the sides AB, BC, and CA, respectively. The triangle ABC can be covered by 5 equilateral triangles of side length s. Then, in particular, the six points A, B, C, D, E, and F are also covered by those 5 triangles. By the pigeonhole principle, at least two of these points are covered by the same equilateral triangle of side length s. Therefore, $s \geq \frac{1}{2}$. Since $s \geq \frac{1}{2}$, the triangle ABC can also be covered by only 4 equilateral triangles of side length s; just consider the four equilateral triangles ADF, DBE, ECF, and DEF of side length $\frac{1}{2}$. □

Question 9. A 6×6 square is tiled by 2×1 dominoes that can be placed horizontally or vertically. Must there exist a line cutting the square without cutting any domino?

Answer: Yes, there always exists such a line! We will prove this by contradiction.

Suppose to the contrary that there exists no line that cuts the square without cutting any domino. Then, each of the five horizontal lines cutting the square board must cut at least one (vertical) domino, and each of the five vertical lines cutting the square board must cut at least one (horizontal) domino. Every horizontal line cutting the 6×6 square board partitions it into $6 \times k$ and $6 \times (6-k)$ rectangle boards, $1 \leq k \leq 5$, each containing an even number of 1×1 squares. Thus, any horizontal line cutting the square board and at least one 1×1 vertical domino must cut an even number of vertical dominoes. Moreover, each horizontal line cutting the square board and at least one vertical domino actually cuts at least two vertical dominoes.

Similarly, there must be at least ten distinct horizontal dominoes cut by the five vertical lines cutting the square board. Note that the vertical dominoes cut by one horizontal line cannot be cut by another horizontal line. Therefore, there must be at least ten distinct vertical dominoes cut by the five horizontal lines cutting the square board.

Therefore, there must be at least 20 distinct dominoes tiling the 6×6 square board, which is clearly a contradiction, since there are only 18 dominoes tiling the 6×6 square board. □

Question 10. In this game, all the players start by sitting at a round table, each with a coin in their hand. At every turn, all the players simultaneously put their coin on the table in front of them. Each player can choose the side of the coin that will face down. Then, each player looks at the coins of his left and right neighbors. If the two coins show different faces of the coin, the player in the middle survives for the next turn. Otherwise, if both coins show the same face of the coin, the player in the middle lost.

After everybody knows who is continuing to the next turn, the losers pick up their coins and leave the table. The game is won when there is exactly one player left sitting at the table, who will be the winner and grab the lucky $10 bonus. If all the players are eliminated at the same time, the game is a tie and nobody gets the bonus. If there are only two players left, the players will necessarily eliminate each other at the next turn since left and right neighbors will be the same opponent.

You are offered a chance to play this game and grab the lucky $10 bonus. You can choose to play in a game starting with 10, 11, or 12 players, including yourself. What would you choose and why?

Answer: It does not matter what you choose! No player can win this game, regardless of the number of players,

2.2. LOGICAL CONUNDRUMS

$n \geq 2$, at the beginning! We prove this by showing that the number of players that move onto the next round is always even.

Given $n \geq 2$ players, labeled 1 through n, at the beginning of a round, let $X_i = 1$ if player i selects a head, and 0 otherwise, for $i = 1 : n$. Also, let $S_i = 1$ if player i survives for the next round, and 0 otherwise. Then,

$$S_i = X_{i-1} + X_{i+1} \pmod 2,$$

for $i = 1 : n$, where we used the notations $X_{n+1} \equiv X_1$ and $X_0 \equiv X_n$. Then,

$$\sum_{i=1}^{n} S_i = \sum_{i=1}^{n} (X_{i-1} + X_{i+1}) \pmod 2.$$

Observe that in the last sum each of the terms X_1, X_2, ..., X_n appears exactly wice. Therefore,

$$\sum_{i=1}^{n} S_i = 2 \sum_{i=1}^{n} X_i \pmod 2$$
$$= 0.$$

Thus, the number of S_i's with $S_i = 1$, that is, the number of surviving players for the next round, is even and we will never have a winner since in that case the sum of the numbers S_i in the last round would have to be equal to 1, an odd number. □

Question 11. You are one of n people standing in a circle. Someone outside the circle goes around clockwise and repeatedly eliminates every other person in the circle, until one person – the winner – remains. Where should you stand to become the winner?

Answer: Label the people P_1 through P_n going clockwise around the circle.

First, we prove that when n is a power of 2, $n = 2^k$, $k \in \mathbb{N}$, the winner is P_1. Our proof is by induction on k. For $k = 1$, that is, for two people standing in a circle, P_2 is eliminated and P_1 remains as the winner. Assume that the claim is true for some $k \in \mathbb{N}$, and let $n = 2^{k+1}$ people stand in a circle. After 2^k even–numbered people $\{P_m \mid m \text{ even}\}$ are eliminated, exactly 2^k odd–numbered people $\{P_m \mid m \text{ odd}\}$ remain, so by induction hypothesis, P_1 remains as the winner eventually, hence proving the claim for $k + 1$. This completes the induction and proves our claim for every $k \in \mathbb{N}$.

If $n = 2^k + m$, where $0 \leq m < 2^k$, people P_2, P_4, P_6, ..., P_{2m} are the first m people being eliminated. Once they are eliminated, we are left with 2^k people standing in the circle, and the first person in that circle, namely, person P_{2m+1} will be the winner eventually, by the claim proved in the previous paragraph.

Therefore, person P_{2m+1} is the winner, where $m = n - 2^k$, and 2^k is the maximum power of 2 not exceeding n, that is, $2^k \leq n < 2^{n+1}$. Note that there is an alternative way to present the final result. Given the binary representation of n, $n = \overline{d_1 d_2 d_3 \ldots d_k}$, person P_w is the winner, where the binary representation of w is $w = \overline{d_2 d_3 \ldots d_k d_1}$. □

Question 12. Four pebbles are initially placed on the ground so that they form a square. At each move, you can take an existing pebble from some point P and move it to a new point Q, as long as there is another pebble at the midpoint of PQ. Is it possible to form a larger square using the four pebbles after a finite number of moves?

Answer: No! Suppose the four pebbles are initially placed at the points with coordinates $(0,0)$, $(0,1)$, $(1,0)$ and $(1,1)$. First, observe that, regardless of the (number of) moves, the pebbles are always at points with integer coordinates. Next, note that each move can be reversed, in

2.2. LOGICAL CONUNDRUMS

other words, the reverse of every possible move is also a valid move. So, if it were possible to form a larger square using the four pebbles after a finite number of moves, then these moves reversed and performed in reverse order would lead to a smaller square. However, there is no square, with corners at points with integer coordinates, smaller than the initial square. $\square$

Question 13. An evil troll once captured a bunch of gnomes and told them: "Tomorrow, I will make you stand in a file, one behind the other, ordered by height such that the tallest gnome can see everybody in front of him. I will place either a white cap or a black cap on each head. Then, starting from the tallest, each gnome has to declare aloud what he thinks the color of his own cap is. In the end, those who were correct will be spared; the others will be eaten, silently."

The gnomes thought about it and came up with an optimal strategy. How many of them survived? What if the hats come in 10 different colors?

Answer: We will show that there is a strategy that saves everyone except for the tallest gnome. Observe that the problem is equivalent to the one in which gnomes hats have numbers instead of colors. If there are k different colors, then the gnomes should replace them with the numbers 0, 1, ..., $k-1$. If there are only two colors, the gnomes should replace them with the numbers 0 and 1. The tallest gnome should sum up all the numbers, calculate its remainder when divided by k, and announce the result to the evil troll. The other gnomes will hear the remainder of the total sum. The second to the tallest gnome will be able to figure out the number that correspond to his hat. After saying that number, the next gnome (third tallest) is in an equivalent position to the second tallest gnome. He can see all the hats in front of him, he has heard the total sum, and he has heard the number on the hat from the gnome behind him. This third tallest gnome can correctly say the number on his hat. The next in line does the same. The gnomes continue with this strategy and all of them are saved, except for the tallest one. □

Question 14. Two competitors play badminton. They play two games, each winning one of them. They then play a third game to determine the overall winner of the match. The winner of a game of badminton is the first player to score at least 21 points with a lead of at least 2 points over the other player. In this particular match, it is observed that the scores of each player listed in order of the games form an arithmetic progression with a nonzero common difference. What are the scores of the two players in the third game?

Answer: Let Alice be the winner of the match; we call the other player Bob. Let Alice's scores in the three games played be a, $a + \alpha$, $a + 2\alpha$, for some $a > 0$, $\alpha \neq 0$. Let Bob's scores in the three games played be b, $b + \beta$, $b + 2\beta$, for some $b > 0$, $\beta \neq 0$.

Firstly, we show that Bob must win the first game. Otherwise, the winners of the three games played would be Alice, Bob, and Alice in that order. In that case,

$$a > b, \ a + \alpha < b + \beta, \ a + 2\alpha > b + 2\beta.$$

Then, $a > b$ and $a + \alpha < b + \beta$ imply that

$$\alpha - \beta \ < \ b - a \ < \ 0.$$

Moreover, $a + 2\alpha > b + 2\beta$ and $a + \alpha < b + \beta$ imply that

$$\alpha - \beta \ > \ \frac{b - a}{2} \ > \ \frac{\alpha - \beta}{2}$$

and therefore $\alpha - \beta > 0$, which is a contradiction.

2.2. LOGICAL CONUNDRUMS

Thus, the winners of the three games played are Bob, Alice, and Alice in that order, and therefore

$$a < b, \ a + \alpha > b + \beta, \ a + 2\alpha > b + 2\beta.$$

From $a < b$ and $a + \alpha > b + \beta$, it follows that

$$\alpha - \beta \ > \ b - a \ > \ 0. \tag{2.24}$$

Moreover, since Alice wins the second game by at least two points, we must have that

$$(a + \alpha) - (b + \beta) \ \geq \ 2,$$

and therefore

$$(a - b) + (\alpha - \beta) \ \geq \ 2. \tag{2.25}$$

From (2.24) and (2.25), we obtain that

$$\begin{aligned}(a + 2\alpha) - (b + 2\beta) &= (a - b) + 2(\alpha - \beta) \\ &> (a - b) + (\alpha - \beta) \\ &\geq 2.\end{aligned}$$

In other words, Alice wins the third game by more than 2 points, which means that her score in the third game must be 21 points. Since $\alpha \neq 0$, her scores across the three games must be decreasing, otherwise Alice could not have win the second game. Thus, $\alpha < 0$ and Alice scored more than 21 points in each of the first two games. Then, it follows that Bob won the first game by exactly two points, which means that $b - a = 2$ and also that Alice won the second game by exactly two points, . which means that $(a + \alpha) - (b + \beta) = 2$. This gives us that $\alpha - \beta = 4$.

We can now see that Alice won the third game by 6 points:

$$(a + 2\alpha) - (b + 2\beta) \ = \ 2(\alpha - \beta) - (b - a) \ = \ 6.$$

Since Alice scored 21 points in the third game, we conclude that the score in the third game was 21 − 15.

For example, the scores in the three games could be as follows:

	Alice Score	Bob Score
Game 1	25	27
Game 2	23	21
Game 3	21	15

More generally, if the negative value of α is denoted by $-k$, where k is a positive integer, that is, if $\alpha = -k$, then the scores in the three games must be as follows:

	Alice Score	Bob Score
Game 1	$21 + 2k$	$23 + 2k$
Game 2	$21 + k$	$23 + k - 4$
Game 3	21	15

(The top table corresponds to $k = 2$.) □

Question 15. There are 25 people sitting around a table and each person has two cards. One of the numbers $1, 2, \ldots, 25$ is written on each card, and each number appears on exactly two cards. At a signal, each person passes one of her cards, the one with the smaller number, to her right hand neighbor. Is it true that at some point one of the players will have two cards with the same numbers?

Answer: We will prove that at some point one of the players must hold two cards with the same number!

Suppose on the contrary that no player ever holds two cards with the same number. Then the two players who each hold a 25 card will never pass those cards and, thus,

2.2. LOGICAL CONUNDRUMS

will hold them forever. Also, a 24 card can only be passed by one of the two players holding a 25 card. So, each 24 card can be passed only at most twice, and then will remain stationary afterwards. Similarly, each 23 card can be passed only by one of the players holding a 24 or a 25 card. Therefore, once the two 24 cards have become stationary, each 23 card can be passed at most four times before it becomes stationary as well.

By applying the same reasoning, once the cards labeled $25, 24, \ldots, 25 - (k - 1)$ have become stationary, each of the two cards labeled $25 - k$, for $k = 1, \ldots, 11$, can be passed at most $2k$ times before it becomes stationary as well. After finitely many sounds of a bell, the twenty four cards labeled 14 through 25 will be in the hands of twenty four different players, where they will remain forever. Thereafter, the remaining (25th) player cannot hold a card with a number larger than 13 and, hence, will never pass a 13 card. However, the other players will always pass a 13 card, so both 13 cards must end up in the hands of the 25th player! This contradicts the assumption that no player ever holds two cards with the same number. □

Question 16. Alice can choose an arbitrary polynomial $p(x)$ of any degree with nonnegative integer coefficients. Bob can infer the coefficients of $p(x)$ by only two evaluations as follows: He chooses a real number a and Alice communicates $p(a)$ to him. He then chooses a real number b and Alice communicates $p(b)$ to him. What values of a and b could help *Bob* succeed and how?

Answer: Bob can choose the values $a = 1$ and $b = p(1) + 1$ and recover Alice's secret polynomial! Here is why:

Denote by

$$p(x) = a_0 + a_1 x + a_2 x^2 + \ldots + a_d x^d$$

Alice's polynomial of arbitrary degree with nonnegative integer coefficients. With his first query $a = 1$, Bob learns $p(1)$ which is exactly the sum of the coefficients $\sum_i a_i$ of $p(x)$. Since all the coefficients of $p(x)$ are nonnegative, the value $b = p(1) + 1$ is larger than any single coefficient of $p(x)$, that is $b > a_i$, fotr any $0 \leq i \leq d$.

Note that $p(b)$, Alice's reply to his second query, is the number N_0 given by

$$N_0 = p(b) = a_0 + a_1 b + a_2 b^2 + \ldots + a_d b^d.$$

This means that Bob can read off the coefficients of $p(x)$ from the base b expansion of N_0 as follows: Firstly, $a_0 = N_0 \pmod{b}$. By setting

$$N_1 = \frac{N_0 - a_0}{b} = a_1 + a_2 b + \ldots + a_d b^{d-1},$$

we can obtain $a_1 = N_1 \pmod{b}$. Then, set

$$N_2 = \frac{N_1 - a_1}{b} =$$

By repeating this process until we retrieve all the coefficients, that is, until we obtain N_i equal to zero. $\square$

Remark. Note that any (integer) value of b greater than $p(1)$ can be used for Bob's second query. Also, note that if we allow coefficients of the secret polynomial to be negative, then Bob cannot do better than d queries, where d is the degree of Alice's polynomial. This is because Alice can always answer Bob's $d-1$ queries $x_1, x_2, \ldots, x_{d-1}$ with a zero, leaving Bob still unable to retrieve her polynomial, since all polynomials of the form $p(x) = (x - x_1) \cdot (x - x_2) \cdots (x - x_d)$ are consistent with her replies.

Question 17. I selected 3 positive integers: x, y, z. You can ask me about two linear combinations of these numbers with any coefficients: For example, you give me a, b

2.2. LOGICAL CONUNDRUMS

and c and I tell you the value of $ax + by + cz$. What is your algorithm to find x, y and z?

Answer: At first, it may seem implausible that such algorithm exists, since it appears that we are being asked to solve a system of two linear equations in three variables. However, one cannot forget that x, y and z are positive integers! The coefficients in the second linear combination can depend on the value of the first linear combination. In particular, if you knew the maximum number of digits any of x, y and z can have, say d, you could set $a = 1$, $b = 10^{-(d+1)}$, $c = 10^{-(2d+2)}$ for your second linear combination and you would be able to find x, y and z from its decimal representation.

More specifically, set $a = 1$, $b = 1$, $c = 1$ for the first linear combination. Denoting the result $ax + by + cz = x + y + z$ by D, let $d = \lceil \log_{10} D \rceil$. At that moment, we can infer that $0 < x, y, z < 10^d$. Then, set $a = 1$, $b = 10^{-d}$, $c = 10^{-2d}$ for the second linear combination. Denoting by S the result

$$S = ax + by + cz = x + y \cdot 10^{-d} + z \cdot 10^{-2d},$$

we find x, y and z as

$$\begin{aligned} x &= \lfloor S \rfloor; \\ y &= \lfloor (S - x) \cdot 10^d \rfloor; \\ z &= \lfloor (S - x) \cdot 10^{2d} - y \cdot 10^d \rfloor. \end{aligned}$$

Finally, note that you can actually find any $n \geq 3$ positive integers from only two linear combinations using the same algorithm! □

Question 18. An entry a_{ij} in a matrix is called a saddle point if it is strictly greater than all the entries in the i–th row of the matrix and strictly smaller than all entries in the j–th column of the matrix, or vice-versa. What is the maximum number of saddle points an $n \times n$ matrix can have?

Answer: A saddle point $a_{i,j}$ is said to be of Type 1, if it is strictly greater than all the entries in the i-th row of the matrix and strictly smaller than all entries in the j-th column of the matrix. Similarly, a saddle point $a_{i,j}$ is said to be of Type 2, if it is strictly smaller than all the entries in the i-th row of the matrix and strictly greater than all entries in the j-th column of the matrix.

We claim there can be at most one saddle point of either type! Suppose there is at least one saddle point of Type 1. Since permuting the rows or columns does not affect the saddle points, we can assume, without loss of generality, that $a_{1,1}$ is a saddle point of Type 1. Suppose there is another saddle point $a_{i,j}$ of Type 1. We have $a_{1,1} > a_{1,j} > a_{i,j}$, as well as $a_{1,1} < a_{i,1} < a_{i,j}$, which clearly leads to a contradiction. Therefore, there is at most one saddle point of Type 1. Similarly, there is at most one saddle point of Type 2.

Therefore, the maximum number of saddle points an $n \times n$ matrix can have is at most two. Actually, it equals two, as one can easily construct matrices with exactly two saddle points, for example,

$$\begin{pmatrix} 1 & 18 & 22 & 4 & 15 \\ 18 & 24 & 17 & 6 & 10 \\ 15 & 21 & 10 & 8 & 12 \\ 3 & 10 & 6 & 2 & 8 \\ 10 & 12 & 7 & 3 & 12 \end{pmatrix}$$

with $a_{4,2} = 10$ and $a_{3,4} = 8$ being the saddle points. $\square$

Question 19. 60 ambassadors are invited to a banquet. Every ambassador has at most 29 enemies. Can you seat

2.2. LOGICAL CONUNDRUMS

all the ambassadors at a large round table such that nobody sits next to an enemy?

Answer: Yes! First, seat the ambassadors at a large round table in an arbitrary way. Let P denote the number of enemy pairs of neighboring ambassadors in that seating. We will devise an algorithm which modifies the seating and decreases P, whenever P is positive. Iterating this algorithm eventually achieves $P = 0$, in other words, yields a seating where no ambassador sits next to an enemy. The description of our algorithm is as follows.

Assuming there is at least one enemy pair of neighboring ambassadors, let (A, B) denote one such pair with B sitting to the right of A. Since every ambassador has at most 29 enemies, then every ambassador has at least 30 friends. Starting with ambassador A, go around the table counterclockwise. You encounter at least 30 friends of A. The seats to their right cannot all be occupied by enemies of ambassador B, since ambassador B has at most 29 enemies. Therefore, there is a friend C of the ambassador A whose right neighbor D is a friend of B. Let $X_1, X_2, \ldots, X_k$ be the ambassadors between B and C in counter-clockwise orientation. We change the position of the ambassadors $B, X_1, X_2, \ldots, X_k, C$ into $C, X_k, \ldots, X_2, X_1, B$ in counter-clockwise orientation. After this switch, the friendly ambassadors B and D will sit next to each other, and the friendly ambassadors A and C will sit next to each other. Every other pair of neighbors remain the neighbors in this new seating. This procedure will decrease the number P by at least one. The procedure can be repeated for as long as P is not zero. □

Question 20. You are in a glitzy casino in Las Vegas. Having tried your hand at everything from Roulette to Black Jack, you managed to lose most of your money and have only one dollar left. What's worse, with all the champagne and everything, you misbehaved and the management made it very clear that you are not allowed to play any more games. However, you need two dollars to take the bus back to the hotel. Two shady characters at the bar offer you a game: they have a pile of 15 stones. Each of you in turn is to take your choice of 1, 2, 3, 4 or 5 stones from the pile. The person who takes the last stone gets one dollar from the person who drew previously, and the third person neither wins nor loses. You are to draw first. If both of the other players will play to their best personal advantage and will not make any mistakes, should you agree to play the game?

Answer: We will partition the set of all positive integers into three disjoint subsets: K, K', and K''.

An integer x is an element of K if it satisfies the following property: If the number of stones on the pile is x, then the player who has to take the next stone will eventually lose the game and end up paying \$1, provided that the other two players play to their best interest.

An integer x is an element of K'' if it satisfies the following property: If the number of stones on the pile is x, then the player who has to take the next stone has a winning strategy that will result in him/her receiving \$1, provided that each of the other two players aims to maximize their gain.

An integer x is an element of K' if it satisfies the following property: If the number of stones on the pile is x, then the player who has to take the next stone will not be able to win \$1; however, he/she has a strategy to walk away without loosing money.

Clearly, the numbers 1, 2, 3, 4, and 5 belong to K''. The number 6 belongs to K. We will now prove that the number 7 belongs to K'. The player who sees the pile with 7 stones cannot take all stones and win immediately. If he/she takes ≥ 2 stones from the pile, than the next player

2.2. LOGICAL CONUNDRUMS

will take all the remaining stones. If he/she takes 1 stone, then the next player will face the pile with 6 stones and lose one dollar.

Next, we prove that the numbers 8, 9, 10, 11, and 12 belong to K''. Assume that the number of stones x is an element of the set $\{8, 9, \ldots, 12\}$. Denote by A the player whose turn is next. Denote by B the player who will play immediately after the player A, and by C the player who will play after the player B. The move of the player A should result in exactly 7 stones remaining in the pile. We already established that $7 \in K'$. Therefore, the player B cannot win. The player B can either lose \$1 or walk away with \$0. We assumed that the player B will play in such a way to not lose money. The player B will take exactly one stone. The player C will be facing the pile with 6 stones. No mater what the player C does, the player A can finish the game in the next move.

We will now prove that $13 \in K$. The player who gets to take stones from the pile of size 13, will definitely leave a pile whose number of stones belongs to $\{8, 9, \ldots, 12\} \subseteq K''$. This means that the next player is the winner.

We are ready to prove that $14 \in K'$. Let U be the player who must take stones from the pile whose size is 14. Let V be the player who will play after U and let W be the player who will play after V. Observe that U must not take more than 1 stone from the pile. Indeed, if U takes more than one stone, then U will leave the pile whose number of stones is in the set $\{9, 10, 11, 12\} \subseteq K''$. This would allow the player V to be the winner, which means that U will lose 1 dollar. However, if U takes exactly 1 stone, then V will face the pile with 13 stones. The number 13 belongs to K, which implies that the player V will lose the game if W and U play to their advantage. Hence, the player U has the strategy that will guarantee that U does not lose the money.

We can now prove that

$$\{15, 16, 17, 18, 19\} \subseteq K''.$$

Denote by Q the player who faces the pile whose size belongs to $\{15, 16, 17, 18, 19\}$. Let R be the player that will player after Q, and S the player that will play after R. The player Q should take as many stones as necessary to leave the pile with 14 stones. Then Q will have a choice: either lose \$1 or walk away with nothing. We assumed that Q is a rational player. This means that Q will not lose money, if possible. The player Q can save him/herself from loosing if he/she takes exactly 1 stone and puts the player R in front of the loosing pile of size 13.

Since $15 \in K''$, the game that starts with the pile of 15 stones is the one worth playing. The player who starts the game has the winning strategy. $\square$

Question 21. You have three identical Fabergé eggs, either of which would break if dropped from the top of a 100–floors building. Your task is to determine the highest floor from which the eggs can be dropped without breaking. What is the minimum number of drops required to achieve this? You are allowed to break all the eggs in the process.

Answer: Consider the following more general problem:

Find the largest number of floors $h_e(n)$ a building could have in order to be able to determine the highest floor from which an egg could be dropped without breaking using e eggs and n drops.

Since one drop can only determine one floor, it follows that

$$h_e(1) = 1. \tag{2.26}$$

2.2. LOGICAL CONUNDRUMS

If we have only one egg at our disposal, the only possible strategy is to try the floors one by one from bottom to top; hence,

$$h_1(n) = n. \tag{2.27}$$

When $e \geq 2$ and $n \geq 2$, the first drop cannot be from the floor higher than $h_{e-1}(n-1) + 1$, since if the egg breaks, there are only $e-1$ eggs and $n-1$ drops left, and the highest floor we can still handle is $h_{e-1}(n-1)$. If the first drop does not break an egg, we can treat floor $h_{e-1}(n-1) + 2$ as the new floor 1, and reduce the problem to a problem with e eggs and $n-1$ drops; then,

$$h_e(n) = 1 + h_{e-1}(n-1) + h_e(n-1).$$

We can use the previous formula twice to obtain

$$\begin{aligned} & h_e(n) \\ =\ & 1 + h_{e-1}(n-1) + h_e(n-1) \\ =\ & 2 + h_{e-1}(n-1) + h_{e-1}(n-2) + h_e(n-2). \end{aligned}$$

Iterating this argument gives us

$$\begin{aligned} h_e(n) &= (n-1) + \sum_{j=1}^{n-1} h_{e-1}(j) + h_e(1) \\ &= n + \sum_{j=1}^{n-1} h_{e-1}(j), \end{aligned} \tag{2.28}$$

since $h_e(1) = 1$; see (2.26).

For $e = 2$ and using (2.27), we obtain from (2.28) that

$$\begin{aligned} h_2(n) &= n + \sum_{j=1}^{n-1} h_1(j) \\ &= n + \sum_{j=1}^{n-1} j \ =\ n + \frac{(n-1)n}{2} \tag{2.29} \\ &= \frac{n(n+1)}{2}, \tag{2.30} \end{aligned}$$

where (2.29) follows from the summation formula

$$\sum_{j=1}^{k} j = \frac{k(k+1)}{2};$$

see Appendix 6.1.

Furthermore, for $e = 3$ in (2.28) and using (2.30), we obtain that

$$\begin{aligned} h_3(n) \\ &= n + \sum_{j=1}^{n-1} h_2(j) = n + \sum_{j=1}^{n-1} \frac{j(j+1)}{2} \\ &= n + \frac{1}{2}\sum_{j=1}^{n-1} j + \frac{1}{2}\sum_{j=1}^{n-1} j^2 \\ &= n + \frac{1}{2}\frac{(n-1)n}{2} + \frac{1}{2}\frac{(n-1)n(2n-1)}{6} \quad (2.31) \\ &= \frac{n(n^2+5)}{6}, \quad (2.32) \end{aligned}$$

where (2.31) follows from the summation formulas

$$\sum_{j=1}^{k} j = \frac{k(k+1)}{2}; \quad \sum_{j=1}^{k} j^2 = \frac{k(k+1)(2k+1)}{6};$$

see Appendix 6.1.

Since $h_3(8) = 92 < 100 < h_3(9) = 129$, we conclude that the required number of drops is 9. □

Question 22. A rectangular table has 100 coins placed on it such that none of the coins overlap and it is impossible to put any more coins on the table without causing an overlap. Can you completely cover the table with 400 overlapping coins?

2.2. LOGICAL CONUNDRUMS

Answer: Yes! Without loss of generality, we can assume that the 100 non-overlapping coins, initially on the rectangular table, have unit radius. Since it is impossible to place any more coins on the table without causing an overlap, there is no point P on the table, that is at least distance 2 from each of the coins' centers. Indeed, if there were such a point P on the table, we could place an additional coin centered at P, without overlapping the original 100 coins.

So, if each of the 100 coins of unit radius were replaced by a big coin of radius 2 (and same center), these 100 big coins would completely cover the table, since every point on the table is within distance 2 from the center of (at least) one of the (original) coins. Shrink this arrangement of 100 big coins by a factor of 2 to obtain a covering of, say, the bottom left–hand quadrant of the rectangular table with 100 overlapping coins of unit radius. Do likewise for the other three quadrants of the table and we obtain a covering of the rectangular table with 400 overlapping coins of unit radius. □

Question 23. There are 100 coins on a table: 30 are genuine and 70 are fake. You know that all the genuine coins have the same weight; every fake coin is heavier than any genuine coin; and no two fake coins have the same weight. You have a balance with two pans (but no weights). What is the smallest possible number of weighings to identify at least one genuine coin?

Answer: We can certainly identify at least one genuine coin in 70 weighings: in each weighing, simply weigh the lightest coin you have found so far against a coin that has not been used so far. After 70 weighings, the lightest coin we have found is definitely genuine.

Next, we construct an example of 100 such coins that shows that 70 weighings may be necessary! Suppose the genuine coins all have weight 0, while the fake coins have weights that are distinct powers of 2, say, $2^0, 2^1, \ldots, 2^{69}$. Since $2^0 + 2^1 + \ldots + 2^m = 2^{m+1} - 1$, for every nonnegative integer m, we deduce that

$$\sum_{s \in S} 2^s < 2^{\max_{s \in S} s + 1},$$

for any subset S of $\{0, 1, 2, \ldots, m\}$. Hence, any single weighing, regardless of the coins involved, will simply tip the scale to the side of the heaviest fake coin used, since that coin alone weighs more than the other coins in the opposite pan of the balance altogether. Next, imagine that our balance with two pans has the extra feature that it marks the heaviest (fake) coin whenever a weighing with some coins is performed. We learn no information about any other coins used in that weighing, and there is no point in using the marked coin again in any subsequent weighing. Essentially, even with this extra feature assumed for our balance, the only information we can gain from any single weighing is to eliminate one coin (the heaviest) as being fake. Therefore, (even with the extra feature assumed) 70 weighings are necessary to eliminate all the fake coins and find at least one genuine coin. □

Question 24. Among 10 given coins, some may be real and some may be fake. All the real coins weigh the same. All the fake coins weigh the same, but have a different weight than the real coins. Can you prove or disprove that all ten coins weigh the same in three weighings on a balance scale?

Answer: Indeed, we can! Note that we do not need to determine whether all the coins are genuine or all the coins are fake; we only need to prove or disprove that they all weigh the same.

2.2. LOGICAL CONUNDRUMS

Divide the 10 coins arbitrarily into four piles, each of different size: pile A with 4 coins, pile B with 3 coins, pile C with 2 coins, and pile D with 1 coin. Using the balance scale, perform the following three weighings: B versus $C \cup D$; A versus $B \cup D$; and $A \cup D$ versus $B \cup C$. Here $X \cup Y$ for piles X and Y denotes the union of piles X and Y.

If (at least) one of these weighings results in an unbalanced scale, then not all the coins weigh the same, and we are done.

Otherwise, each of these three weighings results in a balanced scale. Let a denote the number of fake coins in pile A, b the number of fake coins in pile B, c the number of fake coins in pile C, and d the number of fake coins in pile D. Then,

$$\begin{aligned} b &= c + d, \\ a &= b + d, \\ a + d &= b + c. \end{aligned} \qquad (2.33)$$

From (2.33), it follows that $c = 2d$, $b = 3d$, and $a = 4d$. So, $a + b + c + d = 10d$, where d is either 0 or 1, since pile D contains a single coin. Therefore, the total number of fake coins, $a + b + c + d$, is either 0 or 10. In other words, if all three weighings result in a balanced scale, then the coins are either all fake or all genuine; either way, they all weigh the same. $\square$

Question 25. A square grid with 10 rows and 10 columns contains the numbers 1, 2, ..., 100 in its squares written in that order from left to right and from bottom to top. The grid is then tiled with dominos. Each domino is given a value equal to the product of the numbers that are on it. What is the smallest possible sum of the values of the 50 dominos?

Answer: As stated, a domino (a,b) covering numbers a and b is given the value $a \cdot b$. We want to find the minimum value of

$$\sum_{(a,b)} a \cdot b$$

over all possible tilings of the square grid in question. Since $(a-b)^2 = a^2 + b^2 - 2a \cdot b$, it follows that

$$\begin{aligned}
\sum_{(a,b)} (a-b)^2 &= \sum_{(a,b)} \left(a^2 + b^2\right) - 2 \sum_{(a,b)} a \cdot b \\
&= \sum_{i=1}^{100} i^2 - 2 \sum_{(a,b)} a \cdot b \\
&= \frac{100 \cdot 101 \cdot 201}{6} - 2 \sum_{(a,b)} a \cdot b, \quad (2.34)
\end{aligned}$$

where for (2.34) we used the fact that

$$\sum_{i=1}^{n} i^2 = \frac{n(n+1)(2n+1)}{6}$$

for any positive integer n; see (6.2) in Section 6.1.

The key observation is that for a domino (a,b), either $|a-b| = 1$ (if the domino is horizontal), or $|a-b| = 10$ (if the domino is vertical). Therefore, the left-hand side in (2.34) satisfies

$$\sum_{(a,b)} (a-b)^2 \leq 50 \cdot 100. \quad (2.35)$$

The equality is attained when all the dominoes are placed vertically. By combining (2.34) with (2.35), we obtain that

$$\frac{100 \cdot 101 \cdot 201}{6} - 2 \sum_{(a,b)} a \cdot b \leq 50 \cdot 100,$$

2.2. LOGICAL CONUNDRUMS

which is equivalent to

$$\sum_{(a,b)} a \cdot b \geq \frac{1}{2}\left(\frac{100 \cdot 101 \cdot 201}{6} - 50 \cdot 100\right) = 166,675.$$

We conclude that the smallest possible sum of the values of the 50 dominos is $166,675$, attained when all the dominoes are placed vertically. □

Question 26. Four players sit in a circle on chairs numbered clockwise from 1 to 4. Each player has two hats, one black and one white, and is wearing one and holding the other. In the center sits a fifth player who is blindfolded. That player designates the chair numbers of those whose hats should be changed. His goal is to get all four wearing a hat of the same color, in which case the game stops. Otherwise, after each guess, the four walk clockwise past an arbitrary number of chairs (maintaining the same cyclic order), then sit for the next guess. Find a strategy that always works for the blindfolded player.

Answer: The player in the center of the room will only give three kinds of requests to the players on the circle; we call them R_1, R_2, and R_3.

These three requests are defined below and we also specify the sequence in which these three requests should be given:

• The request R_1 asks the players in the chairs 1 and 3 to change the colors of their hats.

• The request R_2 asks the players in the chairs 1 and 2 to change the colors of their hats.

• The request R_3 asks only the player in the chair 1 to change his/her hat.

Denote the white hats by 1 and black hats by -1. Note that the requests R_1 and R_2 preserve the product of all the numbers, while the request R_3 changes the product of the numbers. There are 2^4 possible configurations of hats. Half of them have the product equal to 1 and half of them have the product equal to -1. Denote by S_+ the set of all possible configurations that have the product equal to 1 and by S_- the set of all configurations that have the product equal to -1.

We will now define the sequence Σ_+ of requests whose terms are R_1 and R_2 only. During the requests in Σ_+, the configuration of hats will not move from S_+ to S_- (or vice versa). The sequence Σ_+ will be chosen in such a way to guarantee that at some point all hats become of the same color, provided that the original configuration belongs to S_+. If during the application of the sequence Σ_+ the hats never become of the same color, then we are sure that the original configuration is S_-. In that case, the player in the center of the room gives the request R_3 and repeats the sequence Σ_+. This repeat application of Σ_+ will guarantee that at some point the hats all become of the same color. Thus, once we construct Σ_+, the desired sequence of operations will be

$$(\Sigma_+, R_3, \Sigma_+).$$

In making Σ_+, our goal is to arrange the requests R_1 and R_2 in such a way that if the configuration of hats is in S_+, then at some point all the hats become of the same color. We will thus only work with configurations in S_+. It suffices to restrict our attention only to the subset S_+^2 of those configurations that have two terms equal to 1 and two terms equal to -1. Configurations in S_+ that are not in S_+^2 are trivial: the hats are all of the same color and the game is over. We partition the set S_+^2 in two subsets A and B. The subset A consists of those configurations in which the signs alternate:

$$A = \{(1, -1, 1, -1), (-1, 1, -1, 1)\}.$$

The set B consists of the remaining four configurations.

$$\begin{aligned} B &= \{(1,1,-1,-1),(1,-1,-1,1), \\ &\quad (-1,-1,1,1),(-1,1,1,-1)\}. \end{aligned}$$

Let us now analyze the request R_1. The request R_1 takes a configuration from B into a configuration from B. However, R_1 will take a configuration from A into the one in which all hats are of the same color.

The first term of the sequence Σ_+ will be R_1. If the configuration was in A, then the game is over after this request R_1. If the game is not over yet, then we are sure that the configuration is in B.

Assume now that the request R_2 is given and that the configuration of hats is in B. Since the set B has only 4 elements, it is easy to verify directly that R_2 applied to any element in B will either result in all hats being of the same color, or in all hats falling into a configuration that belongs to A.

Therefore, the sequence Σ_+ defined as

$$\Sigma_+ = (R_1, R_2, R_1)$$

guarantees that at some point all hats are of the same color, provided that initially an even number of hats is black.

Thus, the sequence of requests that, for each initial configuration, guarantees that the hats will be of the same color at some point is

$$(R_1, R_2, R_1, R_3, R_1, R_2, R_1).\quad \square$$

Question 27. There are 100 prisoners in solitary cells. There is a central room with one light bulb that is initially off. No prisoners can see the light bulb from their cells.

Every day, the warden picks a prisoner at random and that prisoner visits the central room. While there, the prisoner can toggle the bulb if he or she wishes. Also, the prisoner has the option of asserting that all 100 prisoners have been to the central room by now. If this assertion is false, all 100 prisoners die. However, if it is indeed true, all the prisoners are set free. The prisoners are allowed to get together one night in the courtyard to discuss their strategy. What strategy should they agree on so that eventually someone can make a correct assertion? What is the expected time and variance of the time until they are all free?

Answer: In order for the prisoners to win the game and get out of the prison alive, they should designate one of them to be the Boss. Every prisoner, other than the Boss, will always be silent. Once the Boss figures out that everybody was in the room at least once, then the Boss will talk to the Warden and make everyone go free. Thus, the behavior of the Boss will be different than the behavior of all other prisoners. If you are a prisoner who is not the Boss, then you have only one goal: make sure that the Boss can figure out if you visited the room already.

Here is the precise strategy that should be used by every prisoner who is not the Boss:

If the switch is in the off position and if you have never touched the switch, then flip the switch from off to on. In every other case, just leave the room and do not touch the switch (i.e. if the switch is on, then just leave the room; if the switch is off, but you have already made one flip from off to on, then don't touch anything, just leave the room).

The Boss will have a different role in the game. The Boss will only flip the switch from on to off and count the number of flips that he/she is making. Once the Boss flipped the switch 99 times, the Boss is certain that everybody was in the room at least once and can say the sentence.

2.2. LOGICAL CONUNDRUMS

Let N be the random variable that denotes the total number of times that the prisoners visited the room with the switch. We need to determine the expected value and variance of N. We will express the random variable N as the sum

$$N = Y_1 + B_1 + Y_2 + B_2 + \cdots + Y_{99} + B_{99} \qquad (2.36)$$

of conveniently chosen random variables Y_1, B_1, Y_2, B_2, ..., Y_{99}, B_{99}. We will design Y_i and B_i, $1 \leq i \leq 99$ in such a way that they are mutually independent and satisfy (2.36).

The random variables Y_1, B_1, Y_2, B_2, ..., Y_{99}, B_{99} will be the lengths of time intervals between the flips of the switch. The time Y_1 is the length of the first time interval during which the position of the switch was off. This first interval ends when a first non-Boss enters the room. The random variable B_1 is the length of the time that the switch was in the on position. This time interval is terminated by the Boss. The random variable Y_2 is the waiting time for the switch to go from off to on. Formally, the random variable Y_i is defined as the length of the i-th interval of the off position of the switch. The random variable B_i is the length of the i-th interval of the on position of the switch. All of these random variables are independent. The random variable B_i has a geometric distribution with parameter $p = \frac{1}{100}$ for every i. The random variable Y_i has a geometric distribution with the parameter $p_i = \frac{100-i}{100}$. The expected value of a Geometric random variable with parameter p is $\frac{1}{p}$ and the variance is $\frac{1-p}{p^2}$. The expected value of N is calculated using the linearity of expectation

$$\begin{aligned}
\mathbb{E}[N] &= \sum_{i=1}^{99} \left(\mathbb{E}[B_i] + \mathbb{E}[Y_i]\right) \\
&= 99 \cdot 100 + \sum_{i=1}^{99} \frac{100}{100-i} \\
&= 9900 + 100 \cdot \left(1 + \frac{1}{2} + \cdots + \frac{1}{99}\right) \\
&\approx 9900 + 100 \cdot (\ln(99) + \gamma) \\
&\approx 1041.73775,
\end{aligned}$$

where $\gamma = 0.57721$ is the Euler constant given by

$$\gamma = \lim_{n \to \infty} \left(\sum_{k=1}^{n} \frac{1}{k} - \ln(n)\right).$$

Since the random variables are independent, it follows from (2.36) that the variance of N is given by

$$\begin{aligned}
\mathrm{var}(N) &= \sum_{i=1}^{99} \left(\mathrm{var}[B_i] + \mathrm{var}[Y_i]\right) \\
&= 99^2 \cdot 100 + 100 \cdot \sum_{i=1}^{100} \frac{i}{(100-i)^2} \\
&\approx 995{,}931. \quad \square
\end{aligned}$$

Question 28. There is a calculator in which all the digits (0 through 9) and the basic arithmetic operators $(+,-,*,/)$ are disabled. However, other scientific functions are operational like *exp*, *log*, *sin*, *cos*, *arctan*, etc. The calculator currently displays a 0. Convert this first to 2 and then to 3.

2.2. LOGICAL CONUNDRUMS

Answer: We will show, in general, how to convert the initial display of 0 to any positive integer n, using this broken calculator.

Let n be a positive integer. Consider the right triangle with legs of length 1 and $\sqrt{n}$. Its hypotenuse has length $\sqrt{n+1}$. Denote by θ the angle across from leg of length $\sqrt{n}$. Then, $\tan \theta = \sqrt{n}$ and

$$\theta = \arctan\left(\sqrt{n}\right).$$

Moreover,

$$\cos \theta = \frac{1}{\sqrt{1+\tan^2 \theta}} = \frac{1}{\sqrt{n+1}}.$$

In other words,

$$\sec \theta = \sec\left(\arctan\left(\sqrt{n}\right)\right) = \sqrt{n+1}. \qquad (2.37)$$

Next, define

$$f(x) = \sec\left(\arctan\left(x\right)\right).$$

Note that $f(x)$ is a function computable on the broken calculator by successively pressing two scientific functions arctan and sec. From (2.37), it follows that

$$f(n) = \sqrt{n+1},$$

for any positive integer n. Then,

$$\begin{aligned} f(0) &= 1; \\ f(f(0)) &= f(1) = \sqrt{2}; \\ f(f(f(0))) &= f(\sqrt{2}) = \sqrt{3}; \\ f(f(f(f(0)))) &= f(\sqrt{3}) = \sqrt{4} = 2. \end{aligned}$$

Hence, we can convert the initial display of 0 to 2 by simply pressing the functions arctan and sec, in that order, four times in a row.

In general, letting f^n denote the n-fold composition

$$f^n = \underbrace{f \circ f \circ \cdots \circ f}_{n \text{ times}},$$

we have that, for every positive integer n,

$$f^n(0) = \sqrt{n} \quad \text{and} f^{n^2}(0) = n.$$

In particular, $f^9(0) = 3$. □

Question 29. Alice and Bob take turns picking up toothpicks off the floor. Alice goes first and she can pick up as many as she wants, but not all of them. In each subsequent turn, the person can pick up any number of toothpicks, as long as it does not exceed the number of toothpicks picked up in the previous turn. Passing your turn is not allowed. The person who picks up the last toothpick wins. Who will win?

Answer: If there is one toothpick on the floor, Alice wins by picking it up on her first move. So, let $n \geq 2$ denote the number of toothpicks on the floor. We will prove that Alice wins if n is not a power of 2; otherwise, Bob wins.

More precisely, we will first prove the following claim by induction: If $n \geq 2$ is not a power of 2, and thus $n = 2^s \cdot t$, for $t \geq 3$, t odd, and $s \geq 0$, then Alice has a winning strategy by picking up 2^s toothpicks on her first move.

The claim is clearly true for the base case $n = 3$ (with $s = 0$, $t = 3$), as Alice wins by picking up $2^s = 2^0 = 1$ toothpick on her first move, Bob picking up 1 toothpick on his first move, and Alice finishing by picking up the last toothpick. Suppose that the claim is true for any number of toothpicks less than n, that is not a power of 2 (the inductive hypothesis). Let $n = 2^s \cdot t$, with $t \geq 3$, t odd, and $s \geq 0$, be the number of toothpicks on the floor,

2.2. LOGICAL CONUNDRUMS

and let Alice pick up 2^s toothpicks on her first move, thus leaving $2^s(t-1)$ toothpicks on the floor. Bob can take at most 2^s toothpicks on his first move, according to the rules.

If Bob picks up 2^s toothpicks on his first move, then there will be $2^s(t-2)$ toothpicks left on the floor. If $t-2=1$, then there are 2^s toothpicks left, so Alice wins by picking up all the remaining toothpicks. Otherwise, since t is odd, $t-2$ is odd and $t-2 \geq 3$, so we apply the inductive hypothesis to the number $2^s(t-2)$ of toothpicks left on the floor, which is clearly less than n and not a power of 2. So, by inductive hypothesis, Alice wins (and her next move is to pick up 2^s toothpicks again).

If Bob picks up $m < 2^s$ toothpicks on his first move, then there will be $2^s(t-1) - m$ toothpicks left on the floor. Note that

$$2^s(t-2) \;<\; 2^s(t-1) - m \;<\; 2^s(t-1),$$

so, being between two multiples of 2^s, the number $2^s(t-1) - m$ of toothpicks left on the floor is not a multiple of 2^s. Hence, $2^s(t-1) - m = 2^{s'} \cdot t'$ with $t' \geq 3$, t' odd, $s' \geq 0$, and $s' < s$. Also, note that $m = 2^s(t-1) - 2^{s'} \cdot t'$ is divisible by $2^{s'}$, in particular, $m \geq 2^{s'}$. Applying the inductive hypothesis to the number $2^s(t-1) - m$ of toothpicks left on the floor, we conclude that Alice wins, by picking up $2^{s'}$ toothpicks on her next move, which is an allowed move since $2^{s'}$ is at most m, the number of toothpicks Bob picked up on his last move.

Next, we prove that Bob wins if $n \geq 2$ is a power of 2, say, $n = 2^s$, for $s \geq 1$. If Alice picks up at least half of all the toothpicks on her first move, then Bob wins by picking up the rest. Otherwise, suppose Alice picks up $m < 2^{s-1}$ toothpicks on her first move, that is, less than half of all the toothpicks. The number of toothpicks left is $2^s - m$, which is not a power of 2, since $2^{s-1} < 2^s - m < 2^s$.

So, $2^s - m = 2^{s'} \cdot t'$ with $t' \geq 3$, t' odd, $s' \geq 0$, and $s' < s$. Also, note that $m = 2^s - 2^{s'} \cdot t'$ is divisible by $2^{s'}$, in particular, $m \geq 2^{s'}$. Now, Bob can adopt Alice's strategy when the number of toothpicks on the floor is not a power of 2, and win by taking $2^{s'}$ toothpicks on his first move, which is an allowed move since $2^{s'}$ is at most m, the number of toothpicks Alice picked up on her last move.

Question 30. A $p \times q \times r$ rectangular box is made from unit cubes. How many unit cubes does a diagonal of the cube go through?

Answer: Let us place the coordinate system in such a way that one corner of the box is at the origin and that the edges are parallel to the axis. Let us consider the diagonal that connects the origin $(0, 0, 0)$ with the opposite vertex (p, q, r). Let us consider a particle that starts at the origin and moves along the diagonal until it reaches the point (p, q, r). We will count how many times this particle leaves one of the unit cubes and enters another. The particle changes the unit cube at exactly those times when at least one of its coordinates is an integer. The x coordinate will be an integer exactly p times. The y coordinate will be an integer exactly q times and the z coordinate will be an integer exactly r times. However, there could be times when two or three of the coordinates are integers simultaneously:

• If two coordinates are integers, then the point is on an edge of a cube.

• If all three coordinates are integers, then the point is a vertex of a cube.

By using the Inclusion–Exclusion Principle, we obtain that the total number of times N that the particle changes the

2.2. LOGICAL CONUNDRUMS

unit cube in which it is located:

$$\begin{aligned}N &= p+q+r-\gcd(p,q)-\gcd(q,r)\\ &\quad -\gcd(r,p)+\gcd(p,q,r),\end{aligned}$$

where $\gcd(a,b)$ is the greatest common divisor of integers a and b. $\square$

Question 31. You are presented with three large buckets. Each of them contains an integer number of ounces of a non–evaporating fluid. At any time you may double the contents of one bucket by pouring into it from a fuller one. Can you always empty one of the buckets eventually?

Answer: Yes! Denote by $a \leq b \leq c$ the initial integer amounts (in ounces) of the liquid in the three large buckets. Doubling the contents of a bucket by pouring into it from a fuller one will be designated as a single *move* in our algorithm. It suffices to devise a *sequence* of moves that transforms the initial (a,b,c) configuration into a configuration (a',b',c') with $\min\{a',b',c'\} < a = \min\{a,b,c\}$. As the minimum of the amounts of liquid in the 3 containers keeps decreasing, repeating this sequence of moves will eventually lead to a configuration with one of the buckets being empty!

We construct the *sequence* of moves as follows. Let $b = q \cdot a + r$, where $0 \leq r < a$. In other words, q is the quotient obtained when b is divided by a, while r is the remainder. Let $q = q_0 \cdot 2^0 + q_1 \cdot 2^1 + \ldots + q_\ell \cdot 2^\ell$ be the unique binary representation of q, with $q_\ell = 1$ being the leading digit, and $q_i \in \{0,1\}$, $i = 0, 1, \ldots, \ell - 1$. The *sequence* consists of $\ell + 1$ steps S_i, $i = 0, 1, \ldots, \ell$. In step S_i: if $q_i = 1$, we pour from B to A, doubling the contents of A; otherwise (if $q_i = 0$), we pour from C to A, doubling the contents of A.

To demonstrate a sequence of moves to the reader, we work out a small example. With $(a,b,c) = (5, 26, 31)$, we have $q = 5$, $r = 1$, $q_0 = 1$, $q_1 = 0$, $q_2 = 1$. Our sequence consists of three steps: S_0, pouring from B to A; S_1, pouring from C to A; and S_2, pouring from B to A. The initial configuration $(5, 26, 31)$ transforms accordingly, as follows:

$$(5, 26, 31) \xrightarrow{S_0} (10, 21, 31) \xrightarrow{S_1} (20, 21, 21) \xrightarrow{S_2} (40, 1, 21),$$

with $1 = b' = \min\{a', b', c'\} < 5 = a = \min\{a, b, c\}$ being the new (smaller) minimum of the amounts of liquid in the three containers.

Note that the amount of liquid in A keeps doubling, so right before step S_i, the bucket A contains $2^i \cdot a$ ounces, $i = 0, 1, \ldots, \ell$. Hence, the total amount of liquid poured from B to A is

$$\sum_{i=0}^{\ell} q_i \cdot 2^i \cdot a = a \cdot \sum_{i=0}^{\ell} q_i \cdot 2^i = a \cdot q.$$

Thus, the amount of liquid left in the bucket B after the sequence of moves is $b' = b - q \cdot a = r < a = \min\{a, b, c\}$.

We still need to make sure that there is enough liquid initially in C to perform the sequence of moves. Note that $q_\ell = 1$, so in the last step, S_ℓ, of the sequence, we definitely pour from B to A. Thus, the total amount of liquid poured from C to A during the sequence of moves is at most

$$\sum_{i=0}^{\ell-1} 2^i a = a \cdot \sum_{i=0}^{\ell-1} 2^i = a \cdot \left(2^\ell - 1\right)$$
$$< a \cdot 2^\ell \leq a \cdot q \leq b \leq c. \quad \square$$

Question 32. You die and awake in hell. Satan awaits you and has prepared a curious game. He arranged n

2.2. LOGICAL CONUNDRUMS

quarters in a line, going in the east/west direction as follows: $THHH\ldots HHHT$. Satan explains the rules of the game: once a day, a coin is removed from the east end, and placed on the west end. If the coin was initially T, you get to choose whether the coin switches to H or not. If the coin was initially H, then Satan gets to make this choice. If at the end of one day all the coins are H, you get to leave Hell; Satan will of course try his hardest to make sure you never leave. Is there a strategy that allows you to eventually leave, or can Satan conspire to keep you in Hell forever?

Answer: We can always leave Hell! In order to describe our strategy, think of each H as a 0-bit and each T as a 1-bit. Then, the n quarters representing the state of the game at any given time actually form the (unique) binary representation of a certain positive integer (the leading 0s, if any, being ignored). For example, for $n = 7$, the initial position $THHHHHT \equiv 1000001$ would represent 65.

Next, consider the game played with Satan in *stages*, each stage consisting of n turns. After each stage, all the quarters are back in their original position, though may not be necessarily showing the same face/bit! In each stage, our strategy is to switch all 1s to 0s, *until* Satan switches a 0 to a 1. Once Satan makes that switch, we stop switching (though Satan may not) and leave the remaining quarters showing $1 \equiv T$ as they are throughout the rest of this stage.

Note that Satan makes at least one switch during every stage. Indeed, otherwise, we would switch all the coins to 0, and leave Hell. Moreover, each stage starts with the digit of the least value and ends with the digit of the largest value. Satan makes the last switch in each stage, switching a 0 to a 1. Therefore, after each stage (consisting of n turns), the number represented in binary increases! After at most 2^n stages (each consisting of n

turns), this numbers reaches 11...11 in binary. At that moment, we switch all 1s to 0s and leave Hell. □

Question 33. If you hang a picture with string looped around two nails, and then remove one of the nails, the picture still hangs around the other nail. Can you come up with a different hanging of the picture with the property that removing *either* nail causes the picture to fall. Finally, can you hang a picture on three nails so that removing any one nail fells the picture?

Answer: The figure below shows how it is possible to hang a picture using two nails such that neither of them can be removed without terrible consequences.

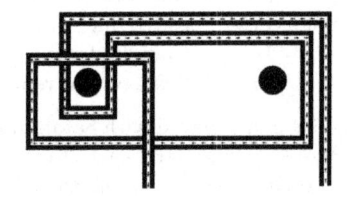

In order to solve an analogous problem with three nails, we will first develop an algebraic representation for the knot from previous picture. Denote by G the non-commutative group whose elements have the form

$$\alpha^{x_1}\beta^{y_1}\alpha^{x_2}\beta^{y_2}\cdots\alpha^{x_n}\beta^{y_n},$$

where $x_1, y_1, \ldots, x_n, y_n$ are integers. We will now show how each configuration of the rope around the nails correspond to one element of the group G. Let e be the unit of the group, i.e., $e = \alpha^0$. The left nail will correspond to α and the right nail to β. Let us set the positive direction to be counter-clockwise. We will follow the rope and write a sentence on the board using the letters α, β,

2.2. LOGICAL CONUNDRUMS

α^{-1}, and β^{-1}. Every time the rope loops around the left nail in positive direction, we write the symbol α on the board. Whenever the string makes a loop around the left nail in the negative direction, we add α^{-1} to the board. The rules for adding β and β^{-1} are analogous. The rope in the picture above corresponds to $\alpha\beta^{-1}\alpha^{-1}\beta$.

We now provide a more formal description of the rules, without writing the entire text on algebraic topology. Without loss of generality, we may assume that the beginning and the ending of the rope are connected. Assume that we place Alice and Bob on top of each of the nails. We place a point at the beginning of the rope (or somewhere at the rope, because now the rope is a circle) and let the point travel along the rope. Alice and Bob make a coordinate system centered at their nails. Let us denote by θ_A the angle that the directional vector of the point on the rope creates with the x axis in the coordinate system that Alice is monitoring. We define the angle θ_B in a similar way.

Whenever Alice notices that the point has made a full circle in positive direction she writes α to the board. The formal definition is this: Alice tracks the value θ_A and when she sees that it became by 2π larger than its former local minimum, she writes α to the board, and resets θ_A to 0. Whenever she notices that the point has made a full circle in negative direction, she adds α^{-1} to the board. Bob performs analogous operations.

It is easy to observe, that in the end, Alice and Bob will end up with the expression $\alpha\beta^{-1}\alpha^{-1}\beta$.

Since the terms α and β do not commute, the expression $\alpha\beta^{-1}\alpha^{-1}\beta$ is not equal to e. However, if we remove the left nail, then we remove every occurrence of α from the expression $\alpha\beta^{-1}\alpha^{-1}\beta$. We obtain $\beta^{-1}\beta = e$, which means that the rope is trivial and not having any nails in its interior. Similarly, if we remove β, we obtain the expression $\alpha\alpha^{-1} = e$.

We are now ready to solve the problem with three nails. Let us consider the free group G generated by α, β, and γ. We need to find an element w of G with the property that if we remove every occurrence of α or every occurrence of β, or every occurrence of γ from w, then the element we obtain becomes e.

Let us denote $\delta = \alpha\beta^{-1}\alpha^{-1}\beta$. We know that, if α or β is removed from δ, we are left with e. The configuration we are looking for is

$$w = \delta\gamma^{-1}\delta^{-1}\gamma.$$

We can see this as follows:

• Removing γ turns w into e.

• Removing α or β from w will first turn every δ into e, which leaves us with $e\gamma^{-1}e\gamma = e$.

• Furthermore, expanding w gives us

$$\begin{aligned} w &= \delta\gamma^{-1}\delta^{-1}\gamma \\ &= \left(\alpha\beta^{-1}\alpha^{-1}\beta\right)\gamma^{-1}\left(\alpha\beta^{-1}\alpha^{-1}\beta\right)^{-1}\gamma \\ &= \alpha\beta^{-1}\alpha^{-1}\beta\gamma^{-1}\beta^{-1}\alpha\beta\alpha^{-1}\gamma. \end{aligned}$$

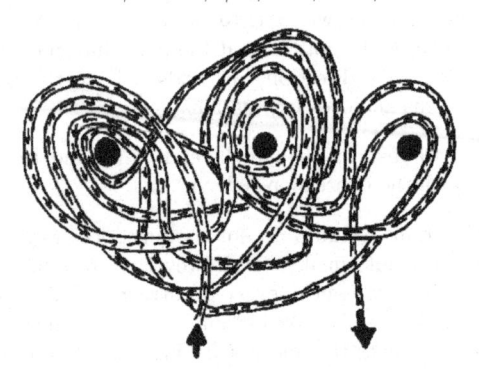

2.2. LOGICAL CONUNDRUMS

The picture above shows the rope that corresponds to w. The removal of any of the nails would cause the picture to fall. □

Question 34. A big rectangle is tiled by finitely many smaller rectangles, each having at least one side of integer length. Show that the big rectangle has at least one side of integer length.

Answer: First Solution: Let A, B, C, and D be the vertices of the big rectangle. Consider the coordinate system in which A is the origin, AB the x-axis, and AD the y-axis. Denote by $\mathcal{S}$ the set of all points E of the plane that are vertices of at least one rectangle. Let $\mathcal{S}'$ be the subset of $\mathcal{S}$ consisting of those points from $\mathcal{S}$ whose both coordinates are integers.

For each point $E \in \mathcal{S}$, denote by n_E the number of rectangles in the partition with one vertex being E. Note that $n_A = n_B = n_C = n_D = 1$. In every other case, the number n_E must be either 2 or 4.

Furthermore, for each rectangle R in the partition, we define $f(R)$ as the number of vertices of R that belong to $\mathcal{S}'$. Since every R has at least one side of integer length, $f(R)$ can take only values 0, 2, or 4. Therefore,

$$\sum_R f(R) \equiv 0 \,(\text{mod } 2).$$

Moreover,

$$\sum_R f(R) = \sum_{E \in \mathcal{S}'} n_E,$$

and therefore

$$\sum_{E \in \mathcal{S}'} n_E \equiv 0 \,(\text{mod } 2).$$

Since A is the origin, we can be sure that $A \in \mathcal{S}'$ and $n_A = 1$. Therefore, at least one other n_E, where $E \in \mathcal{S}'$, must be odd, and that can happen only for E being B, C, or D, since n_E is equal to 2 or 4 otherwise. We are now able to conclude that at least one of the sides of the original rectangle has integral length.

Second Solution: Consider the coordinate system introduced in the first solution. Denote by W the original, big, rectangle. If M is a rectangle whose sides are parallel to the axes of the system, it follows that

$$\int_M \sin(2\pi(x+\alpha))\sin(2\pi(y+\beta))\,dxdy = 0, \quad \forall\, \alpha, \beta \in \mathbb{R}$$

if and only if at least one side of M has integral length. This holds for all rectangles in the partition. By adding up these equalities for each α and β we obtain that

$$\int_W \sin(2\pi(x+\alpha))\sin(2\pi(y+\beta))\,dx\,dy = 0.$$

Thus, W also has a side of integral length. $\square$

Chapter 3

Probabilistic Puzzles

3.1 Discrete Probability

Question 1. There is an unknown total of m misprints in the book and the probability a reader spots a misprint, given that he is looking at it, is a constant. Two readers independently examine the page proofs of a book. Reader 1 flags 30 misprints, while Reader 2 flags 25 misprints. There are 5 flagged misprints in common. Find the number of remaining undetected misprints.

Answer: The total number of misprints flagged by both Reader 1 and Reader 2 is

$$30 + 25 - 5 = 50.$$

Let p denote the probability that Reader 1 spots a misprint, given that he is looking at it. Similarly, let q denote the probability that Reader 2 spots a misprint when he looks at it. Assume that, for each Reader, flagging another misprint is a Bernoulli trial, with probabilities p and q, respectively. Treating the number of misprints flagged by each Reader as the expected value in the corresponding

Bernoulli trial, we obtain

$$mp = 30, \quad mq = 25, \quad mpq = 5.$$

Thus, the total number of misprints in the book can be estimated as

$$m = \frac{(mp) \cdot (mq)}{mpq} = \frac{30 \cdot 25}{5} = 150.$$

Therefore, the number of remaining undetected misprints is $m - 50 = 100$.

Remark. In our attempt to emphasize the idea and intuition, we omitted some important details in the previous solution. Some readers (and interviewers) could request more rigorous solutions. First of all, the problem is not mathematically precise. We are required to make a probabilistic model for the described random experiment. During the construction of the model, we made multiple assumptions. Candidates should ask interviewers to approve the assumptions. The interview questions are quite often not fully precise. They are designed this way with an intention. They want to check whether candidates are able to formally and rigorously describe the assumptions that are being made in order to turn real-world stories into precise and solvable mathematical models.

Our first assumption was that the total number m of misprints is a constant (as opposed to a random variable). Then, we assumed that for each misprint, each reader had a Bernoulli random variable that decided whether the reader detected the misprint. Let $X_1, X_2, \ldots, X_m$ be the Bernoulli random variables that correspond to the Reader 1. For each i, the value of the random variable X_i is equal to 1 if the Reader 1 detected the misprint i. The random variable X_i takes the value 0 if the Reader 1 missed the misprint i. Similarly, let $Y_1, Y_2, \ldots, Y_m$ be the Bernoulli random variables that correspond to the Reader

3.1. DISCRETE PROBABILITY

2. The random variable Y_i has the value 1 or 0 depending on whether the Reader 2 detected the misprint i. From the formulation of the problem we were allowed to assume that the probabilities $\mathbb{P}(X_i = 1)$ do not depend on i. We will denote these probabilities by p. Similarly, the numbers $\mathbb{P}(Y_i = 1)$ do not depend on i and are all equal to q. We made another assumption that was very reasonable, but the problem did not explicitly state that we had the right to make such an assumption. We assumed that the random variables $X_1, Y_1, X_2, Y_2, \ldots, X_m$, and Y_m were independent.

Let X and Y be the random variables that correspond to the number of misprints detected by Readers 1 and 2. Let Z be the total number of misprints detected by both readers. The following equations follow from the definitions of X, Y, and Z:

$$\begin{aligned} X &= X_1 + X_2 + \cdots + X_m, \\ Y &= Y_1 + Y_2 + \cdots + Y_m, \\ Z &= X_1 Y_1 + X_2 Y_2 + \cdots + X_m Y_m. \end{aligned}$$

Due to the linearity of expectations and the assumed independence we have:

$$\begin{aligned} \mathbb{E}[X] &= p \cdot m, \\ \mathbb{E}[Y] &= q \cdot m, \\ \mathbb{E}[Z] &= pq \cdot m. \end{aligned}$$

Now we will describe our last three assumptions. They were $\mathbb{E}[X] = 30$, $\mathbb{E}[Y] = 25$, and $\mathbb{E}[Z] = 5$. Let us provide some justification for making these three assumptions. The objects X, Y, and Z are binomial random variables with realized values 30, 25, and 5. It seems intuitive and reasonable to assume that their expected values are 30, 25, and 5. This can be proved formally using the method of maximal likelihood. However, the rigorous development of the maximal likelihood method is out of scope of this book.

Now, once we have $mp = 30$, $mq = 25$, and $mpq = 5$, we can derive $m = 150$. The number of remaining misprints is 100. □

Question 2. You are lost in a jungle, and you stop at a fork, wanting to know which of two roads lead to the village. Present are three willing natives, one each from a tribe of invariable truth-tellers, a tribe of invariable liars, and a tribe of random answerers. Of course, you don't know which native is from which tribe. You are permitted to ask only two yes-or-no questions, each question being directed to just one native. Can you get the information you need?

Answer: The answer is yes. Our first question will be designed in such a way to identify a person who is definitely not the random answerer. Assume for a moment that we have successfully identified a person who either always tells truth or always lies. We have reduced the problem to the famous standard puzzle in which we are allowed to ask only one question to only one person that will always tell the truth or will always lie. The question to ask is: "If I asked you tomorrow *Is the left road the correct one?*, what would be your answer?" If the answer is yes, then the left road is the correct one. If the answer is no, then the left road is not the correct one. The reasoning is clear in the case in which our conversation was with the truth-teller. If we were dealing with the liar, then our liar was a victim of a double negation. No liar has ever survived one of those.

Let us now focus on designing the *Question 1* whose goal is to eliminate the random answerer. Let A, B, and C be the labels for the three people. We should approach the person A and ask this question:

3.1. DISCRETE PROBABILITY

"If I asked you tomorrow *Is B the random-answerer?*, what would be your answer?"

This will be our *Question 1*. Our strategy should be to follow the advice of the person A. If he/she answered *Yes*, then we should behave as if B is the random answerer and interview the person C with the second question. If the answer is *No*, then the second question should be asked to the person B.

We will now prove that it is safe to believe to the answer to *Question 1*, regardless who the answerer A was.

If A was the truth-teller or the liar, then the logic is the same as before. The truth teller tells the truth because that's what he does, while the liar tells the truth when he is struck with double negation.

However, if the person A is the random-answerer, then either B or C is a good choice for the second question.

Thus, we have constructed the two questions to accomplish these goals: The first questions allows us to identify a person who is not the random answerer. The person who is a consistent liar or a consistent truth-teller is the one that can be successfully interrogated with the second question. □

Question 3. You and your opponent shall play a game with three dice: First, your opponent chooses one of the three dice. Next, you choose one of the remaining two dice. The player who throws the higher number with their chosen dice wins. Now, each dice has three distinct numbers between 1 and 9, with pairs of opposite faces beinng identical. Design the three dice such that you always win! In other words, no matter which dice your opponent chooses, one of the two remaining dice throws a number larger than your opponent, on average.

Answer: Let us consider the following three dice: D_{239}, D_{456}, and D_{178}. The die D_{ijk} has two faces with label i, two faces with label j, and two faces with label k. We will prove the following three propositions

- **Proposition 1.** The die D_{456} on average beats the die D_{239}.
- **Proposition 2.** The die D_{178} on average beats the die D_{456}.
- **Proposition 3.** The die D_{239} on average beats the die D_{178}.

The event that D_{239} beats D_{456} occurs if and only if the roll of the die D_{239} results in 9. The probability of such event is $\frac{1}{3}$. Therefore, with probability $\frac{2}{3}$, the die D_{456} beats the die D_{239}. This proves the Proposition 1.

The event that D_{456} beats D_{178} is equivalent to the event in which the roll of the die D_{178} results in 1. Therefore, the probability that D_{456} beats D_{178} is $\frac{1}{3}$. The probability that D_{178} beats D_{456} is $\frac{2}{3}$, which completes the proof of Proposition 2.

Let us now consider the game in which the dice D_{239} and D_{178} are rolled. The sample space that corresponds to this game is given by the set

$$\Omega \;=\; \{21, 27, 28, 31, 37, 38, 91, 97, 98\}.$$

For $i \in \{2, 3, 9\}$ and $j \in \{1, 7, 8\}$ we use ij to denote the outcome that the roll of D_{239} results in i and the roll of D_{178} results in j. The event B that the die D_{239} beats the die D_{178} satisfies

$$B = \{21, 31, 91, 97, 98.\}.$$

Clearly, $\mathbb{P}(B) = \frac{5}{9} > \frac{1}{2}$. This completes the proof of Proposition 3.

3.1. DISCRETE PROBABILITY

There is a winning strategy for the player who can select the die after seeing which choice the opponent has made. If the first player selects the die D_{456}, then the second player should pick D_{178}. If the first selects the die D_{178}, then the second should select D_{239}. If the first selects D_{239}, then the second player needs to choose the die D_{456}. □

Question 4. Bob chooses a number uniformly at random between 1 and 1000. Alice has to guess the chosen number as quickly as possible. Bob will let Alice know whether her guess is smaller than, larger than or equal to the number. If Alice's guess is smaller than Bob's number, Bob replaces the number with another number chosen uniformly at random from $[1, 1000]$. Prove that there exists a strategy that Alice can use to finish the game in such a way that the expected number of steps is smaller than 45.

Answer: We will use the term *safe strategy* for a strategy that always picks the largest number. If at any point Alice finds out that the Bob's number is in the set $\{1, 2, \ldots, k\}$ and she decides to use the *safe strategy*, then her first guess is the number k. If she is wrong, then her second guess is $k - 1$. If she is wrong again, then her third guess is $k - 2$. With this safe strategy, Bob will never generate a new random number. Alice will keep guessing the numbers until she gets the correct one. Denote by σ_k the expected number of guesses that Alice would make if she uses the safe strategy and if it is certain that the Bob's number is in the set $\{1, 2, \ldots, k\}$. Clearly, we have $\sigma_1 = 1$. Let us prove that, for $k \geq 2$, the following equality holds

$$\sigma_k = 1 + \frac{k-1}{k}\sigma_{k-1}. \qquad (3.1)$$

The equality (3.1) is proved using conditioning on the first roll. Let S_k be the random variable that represents the number of guesses that Alice will make if she uses the safe strategy and the Bob's number is in the set $\{1, \ldots, k\}$. Denote by A the event that Alice was successful in her first guess. The expected value σ_k of the random variable S_k can be written as

$$\begin{aligned} \sigma_k &= \mathbb{E}\left[S_k\right] \\ &= \mathbb{E}\left[S_k|\, A\right] \cdot \mathbb{P}(A) + \mathbb{E}\left[S_k|\, A^C\right] \cdot \mathbb{P}\left(A^C\right). \end{aligned} \quad (3.2)$$

If we condition on the event A, then S_k is a constant equal to 1. This can be written as

$$\mathbb{E}\left[S_k|\, A\right] = 1. \quad (3.3)$$

If we condition on A^C, then the first guess of Alice is definitely a miss. However, the game has changed after this first wrong guess. Alice is now sure that the Bob's number belongs to the set $\{1, \ldots, k-1\}$. Therefore,

$$\mathbb{E}\left[S_k \,\Big|\, A^C\right] = 1 + \sigma_{k-1}. \quad (3.4)$$

The probabilities of the event A and A^C are $\frac{1}{k}$ and $\frac{k-1}{k}$, respectively. Therefore, the equalities (3.2), (3.3), and (3.4) imply (3.1).

By applying formula (3.1) repeatedly, we obtain that

$$\begin{aligned} \sigma_k &= 1 + \frac{k-1}{k}\left(1 + \frac{k-2}{k-1}\sigma_{k-2}\right) \\ &= 1 + \frac{k-1}{k} + \frac{k-2}{k}\sigma_{k-2} \\ &\;\;\vdots \\ &= 1 + \frac{k-1}{k} + \frac{k-2}{k} + \cdots + \frac{1}{k}\sigma_1 \\ &= \frac{1}{k}\left(1 + 2 + \cdots + k\right). \end{aligned}$$

3.1. DISCRETE PROBABILITY

Since $1 + 2 + \cdots + k = \frac{k(k+1)}{2}$, it follows that

$$\sigma_k = \frac{k+1}{2}. \tag{3.5}$$

We will now introduce an improvement to the safe strategy. For given $m \in \{1, \ldots, 1000\}$ we will use the term *safe-m strategy* for the following procedure that consists of two stages:

- **Stage 1.** Alice repeatedly chooses the number m until Bob generates a random numbers that is smaller than or equal to m.

- **Stage 2.** Alice starts applying the *safe strategy* once Bob chooses the number smaller then or equal to m.

The number of guesses that Alice will need in the first stage is a geometric random variable with parameter $\frac{m}{1000}$. Therefore, the expected number of guesses in stage 1 is equal to $\frac{1000}{m}$. The expected number of guesses in stage 2 is $\left(\frac{m+1}{2} - 1\right)$ according to (3.5). We are subtracting 1 from σ_m because the last guess from the stage 1 is included in our calculation that gave us (3.5). Hence, the expected number of guesses under the *safe-m strategy* is

$$f(m) = \frac{1000}{m} + \frac{m-1}{2}.$$

We will now find the value m for which $f(m)$ is minimal. The derivative of $f(x) = \frac{1000}{x} + \frac{x-1}{2}$ is

$$f'(x) = -\frac{1000}{x^2} + \frac{1}{2}.$$

Therefore, the function f is decreasing on $\left(0, \sqrt{2000}\right)$ and increasing on $\left(\sqrt{2000}, +\infty\right)$. Since $\sqrt{2000}$ belongs to the open interval $(44, 45)$, the integer m for which $f(m)$ is minimal must belong to the set $\{44, 45\}$. Direct calculation shows that $f(45) = 44 + \frac{2}{9}$ and $f(44) = 44 + \frac{5}{22}$. The

126   CHAPTER 3. PROBABILISTIC PUZZLES

safe-45 strategy satisfies the requirements from the problem. The *safe*-44 strategy also satisfies the requirements of the problem but is a little bit worse than *safe*-45. $\square$

Question 5. Given a regular 17-gon, choose 3 vertices at random and form a triangle. What is the probability that the center of the polygon is inside the triangle?

Answer: The experiment of selecting the three vertices can be performed in three stages. In stage one we select one vertex and label it by 0. Then we label the remaining vertices with 1, 2, ..., 16 in counter-clockwise orientation. In the second stage we select the vertex A_1 from the set $\{1, 2, \ldots, 16\}$. In the third stage we select the vertex A_2 from the set $\{1, 2, \ldots, 16\} \setminus \{A_1\}$. Denote by G the event that the center of the polygon is inside the triangle. The probability of the event G can be written as

$$\begin{aligned}
\mathbb{P}(G) &= \sum_{i=1}^{16} \mathbb{P}(G \cap \{A_1 = i\}) \\
&= \sum_{i=1}^{16} \mathbb{P}(G \mid A_1 = i) \cdot \mathbb{P}(A_1 = i) \\
&= \frac{1}{16} \sum_{i=1}^{16} \mathbb{P}(G \mid A_1 = i), \quad (3.6)
\end{aligned}$$

since $\mathbb{P}(A_1 = i) = \frac{1}{16}$ for every $i = 1 : 16$. Note that

$$\mathbb{P}(G \mid A_1 = i) = \mathbb{P}(G \mid A_1 = 17 - i),$$

due to symmetry. Then, (3.6) becomes

$$\mathbb{P}(G) = \frac{1}{8} \sum_{i=1}^{8} \mathbb{P}(G \mid A_1 = i). \quad (3.7)$$

Let us fix $i \in \{1, 2, \ldots, 8\}$. The conditional probability $\mathbb{P}(G \mid A_1 = i)$ is equal to the probability that the points

3.1. DISCRETE PROBABILITY

0, i, and A_2 form an acute-angled triangle, where A_2 is a point chosen uniformly at random from the set

$$\{1, 2, \ldots, 16\} \setminus \{i\}.$$

The triangle will be acute-angled if and only if the point A_2 belongs to the set $\{8+1, \ldots, 8+i\}$. This probability is equal to $\frac{i}{15}$ and therefore

$$\mathbb{P}(G \,|\, A_1 = i) \;=\; \frac{i}{15}. \tag{3.8}$$

From (3.7) and (3.8), we obtain that

$$\mathbb{P}(G) \;=\; \frac{1}{8} \sum_{i=1}^{8} \frac{i}{15} \;=\; \frac{1}{8 \cdot 15} \cdot \frac{8 \cdot 9}{2} \;=\; \frac{3}{10}.$$

We conclude that the probability that the triangle contains the center of the polygon is 30%. □

Question 6. Find the probability that, in the process of repeatedly flipping a fair coin, you will encounter a run of 5 heads before a run of 2 tails.

Answer: Let A be the event that a run of 5 heads occurs before a run of 2 tails. Let us denote $\alpha = \mathbb{P}(A)$. We need to determine α. Let F_i be the event that the first i tosses are heads. We will also write F_T instead of F_1^C because it is intuitive: F_1^C means that the first toss was not H, which means that the first toss must have been T. Then,

$$\begin{aligned}\alpha \;&=\; \mathbb{P}(A\,|\,F_1) \cdot \mathbb{P}(F_1) \;+\; \mathbb{P}(A\,|\,F_T) \cdot \mathbb{P}(F_T) \\ &=\; \frac{1}{2}\left(P(A\,|\,F_1) + P(A\,|\,F_T)\right).\end{aligned} \tag{3.9}$$

Let

$$\begin{aligned}\beta \;&=\; \mathbb{P}\left(A\,\big|\,F_1^C\right) \;=\; \mathbb{P}(A\,|\,F_T); \\ \alpha_i \;&=\; \mathbb{P}(A\,|\,F_i), \;\text{ for } i = 1:4.\end{aligned}$$

Then, the equation (3.9) can be written as

$$\alpha = \frac{\alpha_1 + \beta}{2}. \qquad (3.10)$$

Denote by F_{TT} the event that the first two tosses are tails and by F_{TH} the event that the first toss is T and the second toss is H. Then,

$$\begin{aligned}\beta &= \mathbb{P}(A|\,F_{TT}) \cdot \mathbb{P}(F_{TT}|\,F_T) \\ &\quad + \mathbb{P}(A|\,F_{TH}) \cdot \mathbb{P}(F_{TH}|\,F_T). \end{aligned} \qquad (3.11)$$

Note that

$$\mathbb{P}(A|\,F_{TT}) = 0,\ \mathbb{P}(A|\,F_{TH}) = \alpha_1,\ \text{and}$$
$$P(F_{TH}|\,F_T) = \frac{1}{2}.$$

Therefore, (3.11) becomes

$$\beta = \frac{\alpha_1}{2}. \qquad (3.12)$$

A similar reasoning gives us

$$\alpha_i = \frac{\alpha_{i+1} + \beta}{2}, \quad \forall\, i = 1:4. \qquad (3.13)$$

Using (3.13) repeatedly, α_1 can then be fully expressed in terms of β as follows:

$$\begin{aligned}\alpha_1 &= \frac{1}{2}\alpha_2 + \frac{1}{2}\beta = \frac{1}{4}(\alpha_3 + \beta) + \frac{1}{2}\beta \\ &= \frac{1}{4}\alpha_3 + \beta\left(\frac{1}{2} + \frac{1}{4}\right) \\ &= \frac{1}{8}\alpha_4 + \beta\left(\frac{1}{2} + \frac{1}{4} + \frac{1}{8}\right) \\ &= \frac{1}{16}\alpha_5 + \beta\left(\frac{1}{2} + \frac{1}{4} + \frac{1}{8} + \frac{1}{16}\right). \end{aligned}$$

3.1. DISCRETE PROBABILITY

Since $\alpha_5 = \mathbb{P}(A|\, F_5) = 1$, we conclude that

$$\alpha_1 = \frac{1}{16} + \frac{15}{16}\beta. \tag{3.14}$$

From (3.12) and (3.14), we find that $\alpha_1 = \frac{2}{17}$ and $\beta = \frac{1}{17}$. By substituting these two values in (3.10), we obtain that the probability that you will encounter a run of 5 heads before a run of 2 tails is $\alpha = \frac{3}{34}$. $\square$

Question 7. Alice keeps tossing her fair coin until she gets 2 heads in a row. Bob keeps tossing his fair coin until he gets 3 heads in a row. Let A denote the number of Alice's tosses, and B denote the number of Bob's tosses. Find $\mathbb{P}(A > B)$.

Answer: First Solution: We will say that this coin tossing match between Alice and Bob is in state (a, b), $0 \leq a \leq 2$, $0 \leq b \leq 3$, if, when Alice and Bob are about to toss their respective fair coins next, Alice happens to be on a streak of a consecutive heads and Bob happens to be on a streak of b consecutive heads. Define

$$p_{(a,b)} = \mathbb{P}(A > B \mid \text{ match is in state } (a,b)).$$

We are asked to compute $p_{(0,0)}$. Clearly,

$$p_{(0,3)} = 1, \ p_{(1,3)} = 1, \ p_{(2,0)} = 0,$$
$$p_{(2,1)} = 0, \ p_{(2,2)} = 0, \text{ and } p_{(2,3)} = 0.$$

Given that the match is currently in state $(a, b) \in \{(0, 0),$ $(1, 1)$, $(1, 0)$, $(0, 1)$, $(0, 2)$, $(1, 2)\}$, by conditioning on the outcome of the next coin tosses for Alice and Bob, we obtain the following linear system of six equation in six unknowns $p_{(0,0)}$, $p_{(1,1)}$, $p_{(1,0)}$, $p_{(0,1)}$, $p_{(0,2)}$, and $p_{(1,2)}$.

$$p_{(0,0)} = \frac{1}{4}\left(p_{(0,1)} + p_{(1,0)} + p_{(1,1)} + p_{(0,0)}\right)$$

$$p_{(1,1)} = \frac{1}{4}\left(p_{(0,0)} + p_{(0,2)}\right)$$

$$p_{(1,0)} = \frac{1}{4}\left(p_{(0,1)} + p_{(0,0)}\right)$$

$$p_{(0,1)} = \frac{1}{4}\left(p_{(1,2)} + p_{(1,0)} + p_{(0,2)} + p_{(0,0)}\right)$$

$$p_{(0,2)} = \frac{1}{4}\left(2 + p_{(1,0)} + p_{(0,0)}\right)$$

$$p_{(1,2)} = \frac{1}{4}\left(1 + p_{(0,0)}\right)$$

We will explain the reasoning behind the third and the fifth equation; the remaining equations are derived analogously.

Given that the match is currently in state $(1,0)$, the following four scenarios can occur after the next coin tosses, each with probability $\frac{1}{4}$:

a) If Alice tosses a tail and Bob tosses a head, the match goes to state $(0,1)$ and continues from there.

b) If Alice tosses a tail and Bob tosses a tail, the match goes to state $(0,0)$ and continues from there.

c) If Alice tosses a head and Bob tosses a head, the match goes to state $(2,1)$ and ends there, as $A < B$ occurred, that is, $p_{(2,1)} = 0$.

d) If Alice tosses a head and Bob tosses a tail, the match goes to state $(2,0)$ and ends there, as $A < B$ occurred, that is, $p_{(2,0)} = 0$.

Given that the match is currently in state $(0,2)$, the following four scenarios can occur after the next coin tosses, each with probability $\frac{1}{4}$:

3.1. DISCRETE PROBABILITY

a) If Alice tosses a tail and Bob tosses a head, the match goes to state $(0,3)$ and ends there, as $A > B$ occurred, that is, $p_{(0,3)} = 1$.

b) If Alice tosses a head and Bob tosses a head, the match goes to state $(1,3)$ and ends there, as $A > B$ occurred, that is, $p_{(1,3)} = 1$.

c) If Alice tosses a head and Bob tosses a tail, the match goes to state $(1,0)$ and continues from there.

d) If Alice tosses a tail and Bob tosses a tail, the match goes to state $(0,0)$ and continues from there.

Solving this system (by hand or by your favorite linear solver) yields $p_{(0,0)} = \frac{361}{1699}$.

Second Solution: Since $B \geq 3$, the probability of the event $\{A > B\}$ can be written as

$$\begin{aligned} \mathbb{P}(A > B) &= \sum_{k=3}^{\infty} \mathbb{P}(A > B \mid B = k) \mathbb{P}(B = k) \\ &= \sum_{k=3}^{\infty} \mathbb{P}(A > k) \mathbb{P}(B = k). \end{aligned}$$

Let $m \geq 2$ and denote by N_m the number of tosses until m consecutive heads appear. Let

$$\alpha_m(k) = \mathbb{P}(N_m > k).$$

Observe that $\mathbb{P}(N_m > k) = 1$ for all $k < m$. Also, observe that $\mathbb{P}(N_m = m) = \frac{1}{2^m}$ and $\mathbb{P}(N_m > m) = 1 - \frac{1}{2^m}$. Denote by F_i the event that the first i tosses are heads. If k and i are non-negative integers such that $k > m > i$, then we have

$$\begin{aligned} &\mathbb{P}(N_m > k \mid F_i) \\ &= \frac{1}{2} \left(\mathbb{P}(N_m > k \mid F_{i+1}) + \mathbb{P}(N_m > k - i - 1) \right). \end{aligned}$$

The previous equality is obtained by conditioning on the $(i+1)$-st toss. We have used that if the toss $(i+1)$ was tails, then the first $(i+1)$ tosses will not belong to the final sequence of consecutive heads. The waiting time of the remaining game has to be longer than $(k-i-1)$ in order for the waiting time N_m to be longer than k. Therefore,

$$\begin{aligned}\alpha_m(k) &= \mathbb{P}(N_m > k \mid F_0) \\ &= \frac{1}{2}\alpha_m(k-1) + \frac{1}{2}\mathbb{P}(N_m > k \mid F_1) \\ &= \frac{1}{2}\alpha_m(k-1) + \frac{1}{2^2}\alpha_m(k-2) \\ &\quad + \frac{1}{2^2}\mathbb{P}(N_m > k \mid F_3).\end{aligned}$$

Repeating the same procedure and using that the conditional probability $\mathbb{P}(N_m > k \mid F_m)$ is equal to 0 gives us the following formula for $\alpha_m(k)$

$$\begin{aligned}\alpha_m(k) &= \frac{1}{2}\alpha_m(k-1) + \frac{1}{2^2}\alpha_m(k-2) \\ &\quad + \cdots + \frac{1}{2^m}\alpha_m(k-m).\end{aligned} \quad (3.15)$$

Therefore, for fixed m, the sequence $(\alpha_m(k))_{k=1}^{\infty}$ satisfies the linear recursive equation (3.15) with initial values

$$\alpha_m(0) = \alpha_m(1) = \alpha_m(2) = \cdots = \alpha_m(m-1) = 1.$$

Let $\rho_m(1), \rho_m(2), \ldots, \rho_m(m)$ be the complex solutions of the characteristic equation

$$\rho^m - \frac{1}{2}\rho^{m-1} - \frac{1}{2^2}\rho^{m-2} - \cdots - \frac{1}{2^m} = 0.$$

There exist complex numbers $C_m(1), C_m(2), \ldots, C_m(m)$ such that

$$\alpha_m(k) = C_m(1)\rho_m(1)^k + \cdots + C_m(m)\rho_m(m)^k.$$

3.1. DISCRETE PROBABILITY

We will now evaluate $\mathbb{P}(A > B)$ using the identities
$$\mathbb{P}(A > k) = \alpha_2(k) \quad \text{and}$$
$$\mathbb{P}(B = k) = \mathbb{P}(B > k-1) - \mathbb{P}(B > k)$$
$$= \alpha_3(k-1) - \alpha_3(k).$$

The probability of the event $\{A > k\}$ satisfies
$$\mathbb{P}(A > k) = C_2(1)\rho_2^k(1) + C_2(2)\rho_2^k(2).$$

In order to make the expression for $\mathbb{P}(B = k)$ shorter, we introduce the constants $D_3(v)$ for $v \in \{1, 2, 3\}$ in the following way
$$D_3(v) = C_3(v)\left(\frac{1}{\rho_3(v)} - 1\right).$$

The probability of the event $\{B = k\}$ now becomes
$$\mathbb{P}(B = k) = D_3(1)\rho_3^k(1) + D_3(2)\rho_3^k(2) + D_3(3)\rho_3^k(3).$$

We have all the components needed to evaluate the probability $\mathbb{P}(A > B)$.

$$\mathbb{P}(A > B)$$
$$= \sum_{k=3}^{\infty} \left(C_2(1)\rho_2^k(1) + C_2(2)\rho_2^k(2)\right) \times$$
$$\times \left(D_3(1)\rho_3^k(1) + D_3(2)\rho_3^k(2) + D_3(3)\rho_3^k(3)\right)$$
$$= \sum_{u=1}^{2}\sum_{v=1}^{3} \frac{C_2(u)D_3(v)\rho_2^3(u)\rho_3^3(v)}{1 - \rho_2(u)\rho_3(v)}.$$

It remains to evaluate $\rho_2(1)$, $\rho_2(2)$, $\rho_3(1)$, $\rho_3(2)$, $\rho_3(3)$, $C_2(1)$, $C_2(2)$, $C_3(1)$, $C_3(2)$, and $C_3(3)$. The numbers $\rho_2(1)$ and $\rho_2(2)$ are the solutions of the quadratic equation $\rho^2 - \frac{1}{2}\rho - \frac{1}{4} = 0$. Therefore, they are given by

$$\rho_2(1) = \frac{1 - \sqrt{5}}{4},$$
$$\rho_2(2) = \frac{1 + \sqrt{5}}{4}.$$

The numbers $C_2(1)$ and $C_2(2)$ are the solutions of the system

$$C_2(1) + C_2(2) = 1$$
$$C_2(1)\rho_2(1) + C_2(2)\rho_2(2) = 1.$$

The solution of the system is

$$C_2(1) = \frac{5 - 3\sqrt{5}}{10},$$
$$C_2(2) = \frac{5 + 3\sqrt{5}}{10}.$$

The numbers $\rho_3(1)$, $\rho_3(2)$, and $\rho_3(3)$ are the solutions of the equation

$$\rho^3 - \frac{1}{2}\rho^2 - \frac{1}{4}\rho - \frac{1}{8} = 0.$$

Using Cardano's formulas we obtain

$$\rho_3(1) = \frac{1}{6}\left(1 + \sqrt[3]{19 - 3\sqrt{33}} + \sqrt[3]{19 + 3\sqrt{33}}\right),$$
$$\rho_3(2) = \frac{1}{6} - \frac{1}{12}\left(1 + i\sqrt{3}\right)\sqrt[3]{19 - 3\sqrt{33}}$$
$$- \frac{1}{12}\left(1 - i\sqrt{3}\right)\sqrt[3]{19 + 3\sqrt{33}},$$
$$\rho_3(3) = \frac{1}{6} - \frac{1}{12}\left(1 - i\sqrt{3}\right)\sqrt[3]{19 - 3\sqrt{33}}$$
$$- \frac{1}{12}\left(1 + i\sqrt{3}\right)\sqrt[3]{19 + 3\sqrt{33}}.$$

We will use Cramer's rule to obtain $C_3(1)$, $C_3(2)$, and $C_3(3)$ as the solution to the system of equations

$$C_3(1) + C_3(2) + C_3(3) = 1$$
$$C_3(1)\rho_3(1) + C_3(2)\rho_3(2) + C_3(3)\rho_3(3) = 1$$
$$C_3(1)\rho_3^2(1) + C_3(2)\rho_3^2(2) + C_3(3)\rho_3^2(3) = 1.$$

3.1. DISCRETE PROBABILITY

The determinant of the system can be calculated using Vandermonde identity

$$\Delta = \det \begin{vmatrix} 1 & 1 & 1 \\ \rho_3(1) & \rho_3(2) & \rho_3(3) \\ \rho_3^2(1) & \rho_3^2(2) & \rho_3^2(3) \end{vmatrix}$$
$$= (\rho_3(1) - \rho_3(2))(\rho_3(1) - \rho_3(3))(\rho_3(2) - \rho_3(3)).$$

For $v \in \{1, 2, 3\}$, denote by Δ_v the determinant obtained by replacing the v-th column of Δ with the column whose all components are equal to 1. The determinant Δ_v can also be evaluated using Vandermonde identity. The values Δ_1, Δ_2, and Δ_3 are given by

$$\begin{aligned} \Delta_1 &= (1 - \rho_3(2))(1 - \rho_3(3))(\rho_3(2) - \rho_3(3)), \\ \Delta_2 &= (\rho_3(1) - 1)(\rho_3(1) - \rho_3(3))(1 - \rho_3(3)), \\ \Delta_3 &= (\rho_3(1) - \rho_3(2))(\rho_3(1) - 1)(\rho_3(2) - 1). \end{aligned}$$

Cramer's rule now implies

$$\begin{aligned} C_3(1) &= \frac{(1 - \rho_3(2))(1 - \rho_3(3))}{(\rho_3(1) - \rho_3(2))(\rho_3(1) - \rho_3(3))} \\ C_3(2) &= \frac{(\rho_3(1) - 1)(1 - \rho_3(3))}{(\rho_3(1) - \rho_3(2))(\rho_3(2) - \rho_3(3))} \\ C_3(3) &= \frac{(\rho_3(1) - 1)(\rho_3(2) - 1)}{(\rho_3(1) - \rho_3(3))(\rho_3(2) - \rho_3(3))}. \end{aligned}$$

Let us define the constant M in the following way

$$\begin{aligned} M &= \frac{(1 - \rho_3(1))(1 - \rho_3(2))(1 - \rho_3(3))}{\Delta} \\ &= \frac{1^3 - \frac{1}{2} \cdot 1^2 - \frac{1}{4} \cdot 1 - \frac{1}{8}}{\Delta} = \frac{1}{8\Delta}. \end{aligned}$$

Here we used that

$$(x - \rho_3(1))(x - \rho_3(2))(x - \rho_3(3))$$
$$= x^3 - \frac{1}{2}x^2 - \frac{1}{4}x - \frac{1}{8}$$

for every x. The constant $D_3(1)$ can be simplified to

$$\begin{aligned} D_3(1) &= C_3(1)\left(\frac{1}{\rho_3(1)} - 1\right) \\ &= \frac{1-\rho_3(1)}{\rho_3(1)} \cdot \frac{(1-\rho_3(2))(1-\rho_3(3))}{(\rho_3(1)-\rho_3(2))(\rho_3(1)-\rho_3(3))} \\ &= \frac{\rho_3(2) - \rho_3(3)}{\rho_3(1)} \cdot M. \end{aligned}$$

In an analogous way we obtain the equations

$$\begin{aligned} D_3(2) &= \frac{\rho_3(3) - \rho_3(1)}{\rho_3(2)} \cdot M, \\ D_3(3) &= \frac{\rho_3(1) - \rho_3(2)}{\rho_3(3)} \cdot M. \end{aligned}$$

Let us denote $E_1 = \rho_3(2) - \rho_3(3)$, $E_2 = \rho_3(3) - \rho_3(1)$, and $E_3 = \rho_3(1) - \rho_3(2)$. Then we have

$$D_3(v) = \frac{E_v}{\rho_3(v)} M, \quad \text{for } v \in \{1,2,3\}.$$

Let us denote $\xi = \frac{1+i\sqrt{3}}{2}$. Then $\xi^2 = \frac{-1+i\sqrt{3}}{2}$ and ξ solves the equation

$$1 - \xi + \xi^2 = 0.$$

Let us write $\alpha = \sqrt[3]{19 - 3\sqrt{33}}$ and $\beta = \sqrt[3]{19 + 3\sqrt{33}}$. The equations for $\rho_3(1)$, $\rho_3(2)$, and $\rho_3(3)$ can be written in terms of α, β, and ξ.

$$\begin{aligned} \rho_3(1) &= \frac{1 + \alpha + \beta}{6}, \\ \rho_3(2) &= \frac{1 - \xi\alpha + \xi^2\beta}{6}, \\ \rho_3(3) &= \frac{1 + \xi^2\alpha - \xi\beta}{6}. \end{aligned}$$

The determinant Δ satisfies

$$\Delta = \frac{i\sqrt{3}}{72}(\beta^3 - \alpha^3) = \frac{i\sqrt{11}}{4}.$$

3.1. DISCRETE PROBABILITY

From $M = \frac{1}{8\Delta}$ we obtain $M = \frac{-i}{2\sqrt{11}}$. The probability of the event $\{A > B\}$ is

$$\begin{aligned}&\mathbb{P}(A > B)\\&= M \sum_{u=1}^{2} \left(C_2(u)\rho_2^3(u) \cdot \sum_{v=1}^{3} \frac{E_v \rho_3^2(v)}{1 - \rho_2(u)\rho_3(v)} \right)\\&= -\frac{i}{2\sqrt{11}} \sum_{u=1}^{2} \left(C_2(u)\rho_2^3(u) \cdot \sum_{v=1}^{3} \frac{E_v \rho_3^2(v)}{1 - \rho_2(u)\rho_3(v)} \right).\end{aligned}$$

Direct calculations give us that the probability of the event $\{A > B\}$ is approximately equal to 0.21247792. □

Question 8. Two players alternately roll an n-sided fair die. The player who fails to improve upon the previous roll loses. What is the probability that the first player wins?

Answer: First Solution: Denote by D_k the event that the game ends after exactly k rolls. We will prove that the probability of D_k is

$$\mathbb{P}(D_k) = \frac{(k-1) \cdot \binom{n+1}{k}}{n^k}. \tag{3.16}$$

Let ζ_k be the number of sequences $(x_1, x_2, \ldots, x_k)$ of length k such that $x_1 < x_2 < \cdots < x_{k-1}$ and $x_k \leq x_{k-1}$. The probability of D_k is

$$\mathbb{P}(D_k) = \frac{\zeta_k}{n^k}. \tag{3.17}$$

We are going to split ζ_k as

$$\zeta_k = \zeta_k' + \zeta_k'',$$

where ζ_k' denotes the number of sequences $(x_1, x_2, \ldots, x_k)$ of length k such that $x_1 < x_2 < \cdots < x_{k-1}$ and

$x_k \in \{x_1, x_2, \ldots, x_{k-1}\}$, and ζ_k'' denotes the the number of sequences $(x_1, x_2, \ldots, x_k)$ of length k whose terms are all different and for which $x_1 < x_2 < \cdots < x_{k-1}$ and $x_k < x_{k-1}$.

Note that

$$\zeta_k' = (k-1) \cdot \binom{n}{k-1} \tag{3.18}$$

since there are $\binom{n}{k-1}$ ways to choose the terms $x_1, \ldots, x_{k-1}$ and $k-1$ ways to choose the term x_k from the set $\{x_1, x_2, \ldots, x_{k-1}\}$. Similarly,

$$\zeta_k'' = (k-1) \cdot \binom{n}{k}, \tag{3.19}$$

since there are $\binom{n}{k}$ ways to choose k distinct numbers from $\{1, \ldots, n\}$. Once these k numbers are chosen, there are $k-1$ ways to choose x_k. The only requirement is that x_k is not the largest. Once the number x_k is selected the remaining ones must be sorted in increasing order. Thus, $\zeta_k'' = (k-1)\binom{n}{k}$.

From (3.18) and (3.19), we find that

$$\begin{aligned} \zeta_k &= \zeta_k' + \zeta_k'' \\ &= (k-1) \cdot \left(\binom{n}{k-1} + \binom{n}{k} \right) \\ &= (k-1) \cdot \binom{n+1}{k}. \end{aligned} \tag{3.20}$$

The identity (3.16) follows directly from (3.17) and (3.20).

Denote by W the event that the first player wins and by L the event that the first player looses. The events D_k are disjoint. The event W is the union of the events D_k over the even values of k. The event L is the union of the

3.1. DISCRETE PROBABILITY

events D_k over the odd values of k. Thus,

$$\mathbb{P}(W) = \sum_{l \geq 1;\ 2l \leq n+1} (2l-1) \binom{n+1}{2l} \frac{1}{n^{2l}};$$

$$\mathbb{P}(L) = \sum_{l \geq 1;\ 2l+1 \leq n+1} 2l \binom{n+1}{2l+1} \frac{1}{n^{2l+1}}.$$

We define the function f as follows:

$$f(x) = \sum_{k=1}^{\infty} (k-1) \binom{n+1}{k} x^k, \qquad (3.21)$$

where the binomial coefficient $\binom{A}{B}$ is defined to be 0 if $B > A$. Then, $\mathbb{P}(W)$ can be expressed in terms of f as

$$\mathbb{P}(W) = \frac{f\left(\frac{1}{n}\right) + f\left(-\frac{1}{n}\right)}{2}. \qquad (3.22)$$

The function $f(x)$ can be expressed as follows:

$$\begin{aligned}
f(x) &= \sum_{k=1}^{\infty} k \binom{n+1}{k} x^k - \sum_{k=1}^{\infty} \binom{n+1}{k} x^k \\
&= \sum_{k=1}^{\infty} k \cdot \frac{(n+1)!}{k! \cdot (n+1-k)!} x^k - \left((1+x)^{n+1} - 1\right) \\
&= (n+1) \sum_{k=1}^{\infty} \binom{n}{k-1} x^k - \left((1+x)^{n+1} - 1\right) \\
&= (n+1) \cdot x (1+x)^n - (1+x)^{n+1} + 1 \\
&= (1+x)^n (nx-1)) + 1.
\end{aligned}$$

Thus,

$$f\left(\frac{1}{n}\right) = 1; \quad f\left(-\frac{1}{n}\right) = 1 - 2\left(1 - \frac{1}{n}\right)^n. \qquad (3.23)$$

From (3.21) and (3.23), we conclude that the probability that the first player wins is $\mathbb{P}(W) = 1 - \left(1 - \frac{1}{n}\right)^n$.

Second Solution: Denote by p_k the probability that the next player to roll wins the game *given that* the last roll was k, for $k = 1 : n$. Then, $q_k = 1 - p_k$ is the probability that the next player to roll loses the game *given that* the last roll was k. Note that $p_n = 0$ and $q_n = 1$, since the next player to roll cannot improve upon the previous roll, if that roll was n.

Assuming that the last roll in the game was k, $k = 1, 2, \ldots, n$, and conditioning on the outcome of the next roll, we obtain the following equation:

$$\begin{aligned} p_k &= \frac{k}{n} \cdot 0 + \frac{1}{n} \cdot q_{k+1} + \frac{1}{n} \cdot q_{k+2} + \ldots + \frac{1}{n} \cdot q_n \\ &= \frac{1}{n}\left(q_{k+1} + q_{k+2} + \ldots + q_n\right). \end{aligned} \quad (3.24)$$

Indeed, assuming that the last roll in the game was k, if the next player to play rolls a number at most k (which happens with probability $\frac{k}{n}$), he loses the game immediately. Otherwise, he improves upon the previous roll by rolling a number m larger than k, each of the values $m \in \{k+1, k+2, \ldots, n\}$ being equally likely (with probability $\frac{1}{n}$), thus, continuing the game and winning it only if the other player who is next to roll loses the game given that the last roll was m (which happens with probability q_m). By replacing k with $k - 1$ in (3.24), we obtain that

$$p_{k-1} = \frac{1}{n}\left(q_k + q_{k+1} + \ldots + q_n\right), \quad (3.25)$$

for $k = 2 : n$.

By subtracting (3.25) from (3.24), we find that

$$p_k - p_{k-1} = -\frac{1}{n} \cdot q_k,$$

3.1. DISCRETE PROBABILITY

for $k = 2 : n$. Recalling that

$$p_k = 1 - q_k \quad \text{and} \quad p_{k-1} = 1 - q_{k-1},$$

we obtain the recursive relation

$$q_{k-1} - q_k = -\frac{1}{n} \cdot q_k, \ \forall \, k = 2 : n,$$

which can be written as

$$q_{k-1} = \left(1 - \frac{1}{n}\right) q_k, \ \forall \, k = 2 : n. \quad (3.26)$$

By iterating (3.26) and using the fact that $q_n = 1$, it follows that

$$\begin{aligned} q_k &= \left(1 - \frac{1}{n}\right)^{n-k} q_n \\ &= \left(1 - \frac{1}{n}\right)^{n-k}, \ \forall \, k = 1 : n. \quad (3.27) \end{aligned}$$

Finally, denote by p the probability that the first player wins the game. If the first player rolls a number $k \in \{1, 2, \ldots, n\}$ on his first roll (with probability $\frac{1}{n}$), then he will win the game only if the other player next to roll loses the game given that the last roll was k, which happens with probability q_k. Hence,

$$\begin{aligned} p &= \frac{1}{n} \sum_{k=1}^{n} q_k \stackrel{(3.27)}{=} \frac{1}{n} \sum_{k=1}^{n} \left(1 - \frac{1}{n}\right)^{n-k} \\ &\stackrel{i=n-k}{=} \frac{1}{n} \sum_{i=0}^{n-1} \left(1 - \frac{1}{n}\right)^{i} \\ &= \frac{1}{n} \cdot \frac{1 - \left(1 - \frac{1}{n}\right)^n}{1 - \left(1 - \frac{1}{n}\right)} \\ &= 1 - \left(1 - \frac{1}{n}\right)^n. \end{aligned}$$

Therefore, the probability that the first player wins is

$$p = 1 - \left(1 - \frac{1}{n}\right)^n. \quad \Box$$

Question 9. There are 64 teams who play single elimination tournament, hence six rounds, and you have to predict the winners in all 63 games. Your score is then computed as follows: 32 points for correctly predicting the final winner, 16 points for each correct finalist, and so on, down to 1 point for every correctly predicted winner for the first round. (The maximum number of points you can get is thus 192.) Knowing nothing about any team, you flip fair coins to decide every one of your 63 bets. Compute the expected number of points.

Answer: There are 32 games in the first round. Denote by $B_1(1)$, $B_1(2)$, ..., $B_1(32)$ the numbers of points earned from guessing the results in these first 32 games. There are 16 games in the second round. Denote by $B_2(1)$, $B_2(2)$, ..., $B_2(16)$ the numbers of points earned by guessing their results. Similarly, let $B_k(1)$, $B_k(2)$, ..., $B_k(2^{6-k})$ be the numbers of points earned by guessing the results of the games in the round k. The total number of points is

$$N = \sum_{k=1}^{6} \sum_{m=1}^{2^{6-k}} B_k(m).$$

The linearity of expectations implies

$$\mathbb{E}[N] = \sum_{k=1}^{6} \sum_{m=1}^{2^{6-k}} \mathbb{E}[B_k(m)].$$

Observe that for fixed k and m the number $B_k(m)$ is either 2^{k-1} or 0. The value is equal to 2^{k-1} if the winner

3.1. DISCRETE PROBABILITY

is correctly predicted. The number of potential winners in the m-th game of the k-th round is 2^k. Therefore, $\mathbb{E}[B_k(m)] = \frac{1}{2}$. The expected number of points is

$$\begin{aligned} \mathbb{E}[N] &= \sum_{k=1}^{6} \sum_{m=1}^{2^{6-k}} \frac{1}{2} \\ &= \frac{1}{2} \sum_{k=1}^{6} 2^{6-k} \\ &\stackrel{i=6-k}{=} \frac{1}{2} \sum_{i=0}^{5} 2^i \\ &= \frac{63}{2}. \end{aligned}$$

This is consistent with the fact that there are exactly 63 games in the tournament. □

Question 10. Ten percent of the surface of a sphere is colored green and the rest is colored blue. Show that no matter how the colors are arranged, it is possible to inscribe a cube in the sphere so that all its vertices are blue.

Answer: Assume the contrary, that is, assume that it is possible to color 10% of the surface of the sphere in green in such a way that every cube has at least one green vertex. Consider a random cube $A_1 A_2 \ldots A_8$ inscribed in the sphere. Denote by H the random variable defined as the number of green vertices of the cube. Our assumption implies that $H \geq 1$. Therefore $\mathbb{E}[H] \geq 1$.

Let f be the function that assigns the value 1 to green points and the value 0 to blue points. The domain of f is the set of all points of the sphere. The codomain of f is the set $\{0, 1\}$. The random variable H can be expressed in terms of f as follows:

$$H(A_1 A_2 \ldots A_8) = f(A_1) + f(A_2) + \cdots + f(A_8).$$

If X is a point chosen uniformly at random from the sphere, then the expected value of $f(X)$ is exactly 0.1. The linearity of expectations gives us

$$\begin{aligned} \mathbb{E}[H] &= \mathbb{E}[f(A_1)] + \mathbb{E}[f(A_2)] + \cdots + \mathbb{E}[f(A_8)] \\ &= 8 \cdot \mathbb{E}[f(A_1)] = 8 \cdot 0.1 = 0.8. \end{aligned}$$

We obtained that $1 \leq \mathbb{E}[H] = 0.8$. This is a contradiction and therefore there must exist a cube whose all vertices are blue. $\square$

Question 11. A person has in each of his two pockets a box with n matches. Now and then he takes a match from a randomly chosen box until he finds the selected box empty. Find the expectation of the number of remaining matches in the other box.

Answer: Denote by X the number of remaining matches in the other box. We are asked to compute $\mathbb{E}[X]$. Label the two matchboxes A and B. Let E_A^k denote the event that there are k matches in box B when the person chooses to take a match from box A that turns out to be empty. The event E_B^k is defined analogously. By symmetry, $\mathbb{P}(E_A^k) = \mathbb{P}(E_B^k)$. Therefore,

$$\mathbb{P}(X = k) = 2\mathbb{P}(E_A^k) \quad \text{for } 0 \leq k \leq n. \qquad (3.28)$$

Note that the event E_A^k occurs if and only if box A is chosen to draw a match from for the $(n+1)$th time exactly in the step

$$((n+1) + (n-k)) = (2n - k + 1).$$

3.1. DISCRETE PROBABILITY

This happens if box A was chosen n times in the first $(2n - k)$ steps which has a probability of $\binom{2n-k}{n} \cdot \left(\frac{1}{2}\right)^{2n-k}$ to occur, and then chosen again in step $2n - k + 1$, which happens with probability $\frac{1}{2}$. Thus,

$$\mathbb{P}\left(E_A^k\right) = \binom{2n-k}{n} \cdot \left(\frac{1}{2}\right)^{2n-k+1}. \qquad (3.29)$$

From (3.28) and (3.29), it follows that

$$\mathbb{P}\left(X = k\right) = \binom{2n-k}{n} \cdot \frac{1}{2^{2n-k}}.$$

Furthermore, note that

$$\begin{aligned}
\mathbb{E}\left[X\right] &= \sum_{k=0}^{n} k \cdot \mathbb{P}\left(X = k\right) \\
&= \sum_{k=0}^{n} k \binom{2n-k}{n} \cdot \frac{1}{2^{2n-k}}.
\end{aligned}$$

This sum appears to be difficult to express in a simple closed form. With this in mind, for $0 \leq k \leq n-1$, let us compute the ratio

$$\begin{aligned}
\frac{\mathbb{P}\left(X = k+1\right)}{\mathbb{P}\left(X = k\right)} &= \frac{\binom{2n-k-1}{n} \cdot \frac{1}{2^{2n-k-1}}}{\binom{2n-k}{n} \cdot \frac{1}{2^{2n-k}}} \\
&= \frac{\frac{(2n-k-1)!}{n!(n-k-1)!} \cdot \frac{1}{2^{2n-k-1}}}{\frac{(2n-k)!}{n!(n-k)!} \cdot \frac{1}{2^{2n-k}}} \\
&= \frac{2n-2k}{2n-k}.
\end{aligned}$$

The last identity can be rewritten as

$$(2n-k)\mathbb{P}\left(X = k+1\right) = (2n-2k)\mathbb{P}\left(X = k\right),$$

or, in an even more convenient format, as

$$(2n+1)\mathbb{P}(X=k+1) - (k+1)\mathbb{P}(X=k+1)$$
$$= 2n\mathbb{P}(X=k) - 2k\mathbb{P}(X=k).$$

By summing up the last identity over $k = 0 : (n-1)$, we find that

$$(2n+1)\sum_{k=0}^{n-1}\mathbb{P}(X=k+1) \tag{3.30}$$

$$- \sum_{k=0}^{n-1}(k+1)\mathbb{P}(X=k+1) \tag{3.31}$$

$$= 2n\sum_{k=0}^{n-1}\mathbb{P}(X=k) - 2\sum_{k=0}^{n-1}k\mathbb{P}(X=k). \tag{3.32}$$

Note that

$$\sum_{k=0}^{n-1}(k+1)\mathbb{P}(X=k+1) = \mathbb{E}[X]; \tag{3.33}$$

$$\sum_{k=0}^{n-1}k\mathbb{P}(X=k) = \mathbb{E}[X] - n\mathbb{P}(X=n), \tag{3.34}$$

while

$$\sum_{k=0}^{n-1}\mathbb{P}(X=k+1) = 1 - \mathbb{P}(X=0); \tag{3.35}$$

$$\sum_{k=0}^{n-1}\mathbb{P}(X=k) = 1 - \mathbb{P}(X=n). \tag{3.36}$$

By using (3.33–3.36) in (3.30–3.32), we obtain that

$$(2n+1)(1-\mathbb{P}(X=0)) - \mathbb{E}[X] \tag{3.37}$$
$$= 2n(1-\mathbb{P}(X=n)) - 2(\mathbb{E}[X] - n\mathbb{P}(X=n)).$$

3.1. DISCRETE PROBABILITY

Note that

$$\mathbb{P}(X=0) = \binom{2n}{n} \cdot \left(\frac{1}{2}\right)^{2n} \text{ and } \mathbb{P}(X=n) = \left(\frac{1}{2}\right)^n.$$

Then, we can solve (3.37) for $\mathbb{E}[X]$ and obtain that

$$\mathbb{E}[X] = \frac{(2n+1)\binom{2n}{n}}{2^{2n}} - 1. \quad \square$$

Question 12. The numbers $1, 2, \ldots, 2020$ on the x-axis are paired up at random to form 1010 intervals. Find the probability that one of these intervals intersects all the others.

Answer: The problem is equivalent to the following one:

The points A_1, B_1, A_2, B_2, $\ldots$, A_{1010}, B_{1010} are chosen uniformly at random on a unit circle centered at origin. For each $i \in \{1, 2, \ldots, 1010\}$, denote by $\widehat{A_i B_i}$ the arc that does not contain the point $(1, 0)$. What is the probability that there exists an index k such that the arc $\widehat{A_k B_k}$ intersect each of the arcs $\widehat{A_i B_i}$?

Each point A_i uniquely corresponds to an angle $\alpha_i \in [0, 2\pi)$. Similarly, each point B_i uniquely corresponds to an angle $\beta_i \in [0, 2\pi)$. The point X with coordinates $(1, 0)$ corresponds to the angle 0. We may assume that all of α_i and β_i are different and different from 0. Starting from the point X and moving counter-clockwise we identify the first of the arcs $\widehat{A_\sigma B_\sigma}$. In other words, σ is defined as the index k for which $\max\{\alpha_k, \beta_k\}$ is minimal.

Let us denote by P the point from $\{A_\sigma, B_\sigma\}$ that is further away from the point X in counter-clockwise direction. In other words, P is the point from $\{A_\sigma, B_\sigma\}$ that corresponds to the angle $\varphi = \max\{\alpha_\sigma, \beta_\sigma\}$. Denote by P' the point from $\{A_\sigma, B_\sigma\} \setminus \{P\}$. Now starting from P, we move in the clockwise direction. For every pair of points (A_ν, B_ν) other than (A_σ, B_σ) we identify the arc $\overset{\frown}{A_\nu B_\nu}$ that does not contain the point P. We identify the first of the arcs $\overset{\frown}{A_\nu B_\nu}$ which contains the point X. Let us denote by Q the point from the set $\{A_\nu, B_\nu\}$ that is further away from P in the clockwise direction. Denote by Q' the point from $\{A_\nu, B_\nu\} \setminus \{Q\}$.

Starting from the point Q and going counter-clockwise we will meet the point X before P' and the point P' before the point P. However, there are 3 possible cases based on the location of the point Q'.

- **Case 1.** The point Q' is between Q and X.
- **Case 2.** The point Q' is between X and P'.
- **Case 3.** The point Q' is between P' and P.

Each three of the cases above occurs with probability $\frac{1}{3}$. We will now prove that in the case 1 there is no index k such that the arc $\overset{\frown}{A_k B_k}$ intersects every other arc. We will then prove that in the cases 2 and 3, the arc $\overset{\frown}{QQ'}$ intersects each of the arcs $\overset{\frown}{A_i B_i}$.

Assume that Q' is between Q and X. Assume that there is an index k such that $\overset{\frown}{A_k B_k}$ intersects both $\overset{\frown}{PP'}$ and $\overset{\frown}{QQ'}$. Then both of the points A_k and B_k must belong to the arc PQ that contains the point X, which contradicts the definition of the index ν and the point Q. Hence, in the case 1, there is no arc that intersects every other arc.

Let us now consider the cases 2 and 3. The point Q' belongs to the arc XP. Assume that there is an arc $\overset{\frown}{A_m B_m}$ that does not intersect $\overset{\frown}{QQ'}$. Then, both endpoints A_m and B_m must belong to the arc QQ' that contains the point X. However, in that situation the index ν would be equal to m, which is a contradiction.

Thus, the required probability is $\frac{2}{3}$. □

3.1. DISCRETE PROBABILITY

Remark. The above idea is due to Graham Brightwell [1]. The problem was first solved in [3].

Question 13. You roll a fair 6-sided die and sum the numbers on the top face as long as you keep rolling larger and larger numbers on the top face. What is expected value of the sum once you stop rolling?

Answer: The die will be rolled at most 6 times. We can modify the game in such a way that the die is always rolled 6 times, however we stop observing and adding the numbers after the roll that is smaller than or equal to the previous one. For every $i \in \{1, 2, \ldots, 6\}$, denote by X_i the result of the i-th roll and by F_i the random variable that denotes the contribution of the i-th roll to the total sum. The random variable F_i is equal to either X_i or to 0. For example, if the six rolls are

$$X_1 = 2,\ X_2 = 5,\ X_3 = 4\ ,$$
$$X_4 = 2,\ X_5 = 3,\ \text{and}\ X_6 = 5,$$

then $F_1 = 2$, $F_2 = 5$, and

$$F_3 = F_4 = F_5 = F_6 = 0.$$

The last four random variables are equal to 0 because only the first two rolls count towards the sum. We need to calculate the expected value of the sum

$$S = F_1 + F_2 + \cdots + F_6.$$

The linearity of expectations implies that

$$\mathbb{E}[S] = \sum_{k=1}^{6} \mathbb{E}[F_k]. \quad (3.38)$$

Fix $k \in \{1, 2, \ldots, 6\}$. The expected value of F_k is

$$\mathbb{E}[F_k] = \sum_{l=1}^{6} l \cdot \mathbb{P}(F_k = l).$$

Since $\{F_k = l\} = \{X_k = l, X_1 < \cdots < X_k\}$, we obtain

$$\begin{aligned} \mathbb{E}[F_k] &= \sum_{l=1}^{6} l \cdot \mathbb{P}(X_k = l, X_1 < \cdots < X_k) \\ &= \sum_{l=1}^{6} l \cdot \mathbb{P}(X_k = l, X_1 < \cdots < X_{k-1} < l). \end{aligned}$$

Since X_k is independent from X_1, X_2, ..., X_{k-1}, and $\mathbb{P}(X_k = l) = \frac{1}{6}$, we obtain that

$$\begin{aligned} \mathbb{E}[F_k] &= \frac{1}{6} \sum_{l=1}^{6} l \cdot \mathbb{P}(X_1 < X_2 < \cdots < X_{k-1} < l) \\ &= \frac{1}{6} \sum_{l=1}^{6} l \cdot \binom{l-1}{k-1} \cdot \frac{1}{6^{k-1}}. \end{aligned} \quad (3.39)$$

From (3.38) and (3.39), we find that the expected value of S is

$$\begin{aligned} \mathbb{E}[S] &= \sum_{k=1}^{6} \sum_{l=1}^{6} l \cdot \binom{l-1}{k-1} \cdot \frac{1}{6^k} \\ &= \frac{1}{6} \sum_{l=1}^{6} l \cdot \sum_{k=1}^{6} \binom{l-1}{k-1} \cdot \frac{1}{6^{k-1}} \\ &= \frac{1}{6} \sum_{l=1}^{6} l \cdot \sum_{k=1}^{l} \binom{l-1}{k-1} \cdot \frac{1}{6^{k-1}}. \end{aligned} \quad (3.40)$$

The binomial formula implies that

$$\sum_{k=1}^{l} \binom{l-1}{k-1} \cdot \frac{1}{6^{k-1}} = \left(1 + \frac{1}{6}\right)^{l-1} = \left(\frac{7}{6}\right)^{l-1}. \quad (3.41)$$

From (3.40) and (3.41), it follows that

$$\mathbb{E}[S] = \frac{1}{6} \sum_{l=1}^{6} l \left(\frac{7}{6}\right)^{l-1}. \quad (3.42)$$

3.1. DISCRETE PROBABILITY

Let us denote

$$G(x) = \sum_{l=1}^{6} l x^{l-1}.$$

Then,

$$\mathbb{E}[S] = \frac{1}{6} G\left(\frac{7}{6}\right). \tag{3.43}$$

Note that

$$\begin{aligned}
G(x) &= \left(\sum_{l=1}^{6} x^l\right)' = \left(\sum_{l=0}^{6} x^l\right)' = \left(\frac{x^7 - 1}{x - 1}\right)' \\
&= \frac{7x^6(x-1) - (x^7 - 1)}{(x-1)^2} \\
&= \frac{6x^7 - 7x^6 + 1}{(x-1)^2}.
\end{aligned}$$

Then, $G\left(\frac{7}{6}\right) = 36$, and, from (3.43), we conclude that

$$\mathbb{E}[S] = \frac{1}{6} \cdot 36 = 6. \quad \square$$

Question 14. Roll six fair dice six times. Each time note the minimum. Find the expected value of the maximum amongst the six minima.

Answer: For $i = 1 : 6$, denote by M_i the random variable that represents the minimum obtained when the six dice were rolled for the i–th time. Let M be the maximum of the random variables $M_1, \ldots, M_6$. Since M is a random variable whose values are positive integers, its expected value can be calculated using the formula

$$\mathbb{E}[M] = \sum_{k=0}^{\infty} \mathbb{P}(M > k); \tag{3.44}$$

see Question 7 from Section 2.1 Discrete Probability of the book "Probability and Stochastic Calculus Quant Interview Questions" [4].

Since $\mathbb{P}(M > k) = 0$ for $k \geq 6$, formula (3.44) reduces to

$$\begin{aligned} \mathbb{E}[M] &= \sum_{k=0}^{5} \mathbb{P}(M > k) \\ &= \sum_{k=0}^{5} (1 - \mathbb{P}(M \leq k)) \\ &= 6 - \sum_{k=0}^{5} \mathbb{P}(M \leq k). \end{aligned} \qquad (3.45)$$

Since

$$\{M \leq k\} = \bigcap_{i=1}^{6} \{M_i \leq k\}$$

and the random variables M_i, $i = 1:6$, are independent, it follows that

$$\mathbb{P}(M \leq k) = \prod_{i=1}^{6} \mathbb{P}(M_i \leq k) = (\mathbb{P}(M_1 \leq k))^6. \quad (3.46)$$

Here, we used the fact that $\mathbb{P}(M_i \leq k)$ is the same for every $1 \leq i \leq 6$, and therefore $\mathbb{P}(M_i \leq k) = \mathbb{P}(M_1 \leq k)$ for $i = 2:6$.

From (3.45) and (3.46), we obtain that

$$\mathbb{E}[M] = 6 - \sum_{k=0}^{5} (\mathbb{P}(M_1 \leq k))^6. \qquad (3.47)$$

To evaluate $\mathbb{P}(M_1 \leq k)$, let $X_1, X_2, \ldots, X_6$ be the individual outcomes of the rolls of the six dice. Since M_1 is the minimum of the random variables $X_1, \ldots, X_6$, the probability of the event $\{M_1 > k\}$ can be calculated using

3.1. DISCRETE PROBABILITY

the independence of X_1, X_2, ..., X_6. If $k \in \{0, 1, ..., 6\}$, then

$$\begin{aligned} \mathbb{P}(M_1 > k) &= \prod_{i=1}^{6} \mathbb{P}(X_i > k) \\ &= (\mathbb{P}(X_1 > k))^6 \\ &= \left(1 - \frac{k}{6}\right)^6. \end{aligned}$$

Thus,

$$\mathbb{P}(M_1 \leq k) = 1 - \mathbb{P}(M_1 > k) = 1 - \left(1 - \frac{k}{6}\right)^6. \quad (3.48)$$

From (3.47) and (3.48), we conclude that

$$\mathbb{E}[M] = 6 - \sum_{k=0}^{5} \left(1 - \left(1 - \frac{k}{6}\right)^6\right)^6 \approx 2.435743051. \quad \square$$

Question 15. How many times do I have to roll a die on average until I roll the same number six times in a row?

Answer: Let E_n be the expected number of rolls until you roll the same number n times in a row. We need to find E_6. Assume that you have just rolled the same number $n-1$ times in a row, which took E_{n-1} rolls on average. Then, one of the following two outcomes will happen:

• with probability $\frac{1}{6}$, you roll the same number again on the next roll; in this case, you end up rolling the same number n times in a row and it took $E_{n-1} + 1$ rolls to obtain this outcome;

• with probability $\frac{5}{6}$, you roll a different number on the next roll; in this case, you already started on your way to roll the same number n times in a row and you will need $E_{n-1} + E_n$ to achieve this.

Hence, we obtain the following recurrence:

$$E_n = \frac{1}{6}(E_{n-1} + 1) + \frac{5}{6}(E_{n-1} + E_n),$$

which is equivalent to

$$E_n = 6E_{n-1} + 1, \ \forall \, n > 1. \tag{3.49}$$

By iterating (3.49), we obtain that

$$E_n = 1 + 6 + \ldots + 6^{n-2} + 6^{n-1}E_1. \tag{3.50}$$

Since $E_1 = 1$, we obtain from (3.50) that

$$E_n = \sum_{i=0}^{n-1} 6^i = \frac{6^n - 1}{5}.$$

Thus, for $n = 6$, we find that $E_6 = 9331$. $\square$

Question 16. A total of n balls, numbered 1 through n, are placed into n boxes, also numbered 1 through n in such a way that ball i is equally likely to go into any of the boxes $1, 2, \ldots, i$. Find the expected number of empty boxes.

Answer: For $i = 1 : n$, define the indicator random variable X_i as $X_i = 1$, if Box i is empty, and $X_i = 0$ otherwise. Denote by X the number of empty boxes. Then,

$$X = \sum_{i=1}^{n} X_i.$$

and therefore

$$\begin{aligned}
\mathbb{E}[X] &= \sum_{i=1}^{n} \mathbb{E}[X_i] \\
&= \sum_{i=1}^{n} \mathbb{P}(\text{Box } i \text{ is empty}). \tag{3.51}
\end{aligned}$$

3.1. DISCRETE PROBABILITY

Note that the only balls that could end up in Box i are the balls $i, i+1, \ldots, n$. Since the balls are placed into boxes independently, it follows that

$$\mathbb{P}(\text{Box } i \text{ is empty}) = \prod_{k=0}^{n-i} \mathbb{P}(\text{ball } i+k \text{ is not in Box } i). \quad (3.52)$$

Since the ball $i+k$ is equally likely to go into any of the boxes $1, 2, \ldots, i+k$, we find that

$$\begin{aligned} \mathbb{P}(\text{ball } i+k \text{ is not in Box } i) &= 1 - \mathbb{P}(\text{ball } i+k \text{ is in Box } i) \\ &= 1 - \frac{1}{i+k} \\ &= \frac{i+k-1}{i+k}. \end{aligned} \quad (3.53)$$

From (3.52) and (3.53), we obtain that

$$\mathbb{P}(\text{Box } i \text{ is empty}) = \prod_{k=0}^{n-i} \frac{i+k-1}{i+k} = \frac{i-1}{n}. \quad (3.54)$$

By combining (3.54) with (3.51), it follows that

$$\begin{aligned} \mathbb{E}[X] &= \sum_{i=1}^{n} \frac{i-1}{n} \stackrel{j=i-1}{=} \frac{1}{n} \sum_{i=0}^{n-1} i \\ &= \frac{1}{n} \cdot \frac{n(n-1)}{2} = \frac{n-1}{2}. \end{aligned}$$

We conclude that the expected number of empty boxes is $\frac{n-1}{2}$. □

Question 17. What is the expected number of cards of a well-shuffled deck that would need to be turned until three spades appear?

Answer: The standard deck of 52 cards (without joker cards) contains 13 cards of each of the 4 suites, in particular, it contains 13 spades.

The main idea for the solution is to consider these 13 spades as dividers. They divide the remaining 39 cards into 14 blocks. "On average," these blocks "should be" of equal lengths. These average lengths are $\frac{39}{14}$. Thus, when the third spade appears, we have seen three blocks of lengths $\frac{39}{14}$ and three spades. Hence, the average number of cards we need to turn over is $3 \cdot \frac{39}{14} + 3 = \frac{159}{14}$. We will see that this answer is correct. However, we need a more rigoruos justification.

Imagine you add a joker card to your deck, with the joker playing a role of the 14th spade! Once the 53 cards in the deck are well-shuffled, you turn them over one by one, placing them along a circle. The 14 spades (including the joker) split the circle into 5 intervals. Now we can use the symmetry of the circle. By symmetry, the lengths of these intervals are equidistributed. Since their lengths add up to 53, by linearity of expectations, the expected length of each of the 14 intervals is $\frac{53}{14}$.

Now, cut the circle open by removing the joker card. We obtain a linear ordering of the cards, with the cards to the right of the joker card being the cards before the first spade. We conclude that that the expected number of cards of a well-shuffled deck that would need to be turned until three spades appear is $3 \cdot \frac{53}{14} = \frac{159}{14}$. □

Question 18. At each stage, one can either pay $1 and receive a coupon that is equally likely to be any of the n types, or one can stop and receive a final reward of $3j$ if one's current collection of coupons contains exactly j distinct types. The objective is to maximize the expected net return. What is the optimal strategy?

3.1. DISCRETE PROBABILITY

Answer: Assume that we have collected j disctinct types of coupons and we are now deciding whether to make another \$1 payment. Let us assume that we are choosing between the following two options:

- *Option 1.* Stop;
- *Option 2.* Play just one more time. Then, stop gambling after this one last gamble, regardless of the outcome.

Let us analyze the "average gain" that each of the two options offer. In the first option, the gain is 0. We would be able to collect our $3j$ dollars, due to the the coupons we already have. However, we are analyzing the gain of this decision to stop playing. That gain is \$0. In option 2, the average gain is $-1 + \frac{3(n-j)}{n}$. The option 2 has better average return if and only if

$$0 < -1 + \frac{3(n-j)}{n}.$$

The last inequality is equivalent to $j < \frac{2n}{3}$.

The real game is different than the simplified game we analyzed. In the real game we are not required to limit ourselves to "only one more gamble." However, we will see that the strategy is going to be the same: Play again for as long as $j < \frac{2n}{3}$.

Formal and rigorous solution: Denote by $M_n(j)$ the expected payoff of the best possible strategy available to the player if there are n types of coupons and the player's collection already contains j different types of coupons. The player has two options: stop the game and take $3j$ dollars, or pay another dollar to keep playing. If the player keeps playing, then, with probability $\frac{n-j}{n}$, the player will receive a coupon of a kind that was not received before. In that case the player will have $j + 1$ coupons and the best possible strategy would bring additional $M_n(j + 1)$

dollars. However, with probability $\frac{j}{n}$ the player would receive a coupon that is already in the collection. Therefore, the number $M_n(j)$ satisfies the following equation

$$M_n(j) \qquad (3.55)$$
$$= \max\left\{3j,\ \frac{n-j}{n}M_n(j+1) + \frac{j}{n}M_n(j) - 1\right\}.$$

From (3.55), it follows that

$$M_n(j) \geq \frac{n-j}{n}M_n(j+1) + \frac{j}{n}M_n(j) - 1,$$

which is equivalent to

$$M_n(j) \geq M_n(j+1) - \frac{n}{n-j}. \qquad (3.56)$$

From (3.55) and (3.56), we obtain that

$$M_n(j) \geq \max\left\{3j,\ M_n(j+1) - \frac{n}{n-j}\right\}. \qquad (3.57)$$

We will now prove that the inequality (3.57) is, in fact, an equality, i.e.,

$$M_n(j) = \max\left\{3j,\ M_n(j+1) - \frac{n}{n-j}\right\}. \qquad (3.58)$$

Assume the contrary, that the inequality in (3.57) is strict for some pair (n, j). Then,

$$M_n(j) > 3j;$$
$$M_n(j) > -\frac{n}{n-j} + M_n(j+1). \qquad (3.59)$$

From (3.55) and since $M_n(j) > 3j$, it follows that we must have

$$M_n(j) = \frac{n-j}{n}M_n(j+1) + \frac{j}{n}M_n(j) - 1,$$

3.1. DISCRETE PROBABILITY

which is equivalent to

$$M_n(j) = M_n(j+1) - \frac{n}{n-j}.$$

This contradicts the strict inequality (3.59). This completes the proof of (3.58).

We now prove that $M_n(j)$ satisfies

$$M_n(j) = \begin{cases} M_n(j+1) - \frac{n}{n-j}, & \text{if } j \leq \frac{2n}{3}; \\ 3j, & \text{if } j > \frac{2n}{3}. \end{cases} \quad (3.60)$$

We first prove that,

$$\text{if } j \leq \frac{2n}{3}, \text{ then } M_n(j) = -\frac{n}{n-j} + M_n(j+1).$$

It suffices to prove that, if $j \leq \frac{2n}{3}$, then

$$3j \leq M_n(j+1) - \frac{n}{n-j}.$$

We will use the fact that $M_n(j+1) \geq 3(j+1)$. Then,

$$\begin{aligned} M_n(j+1) - \frac{n}{n-j} &\geq 3(j+1) - \frac{n}{n-j} \\ &= 3j + \frac{2n-3j}{n-j} \\ &\geq 3j, \end{aligned} \quad (3.61)$$

where, for (3.61), we used the fact that $j \leq \frac{2n}{3}$ and therefore $2n - 3j \geq 0$.

Moreover, we will prove that,

$$\text{if } j > \frac{2n}{3}, \text{ then } M_n(j) = 3j.$$

To do so, we will use induction backwards: first establish that the statement holds for $j = n$, then for $j = n-1$, etc.

In order to use the principle of mathematical induction we first define the predicate $P(k)$ as

$$P(k) \equiv \left[\left(n - k > \frac{2}{3}n\right) \implies (M_n(n-k) = 3(n-k))\right].$$

The statement $P(0)$ is clearly true, because $M_n(n) = 3n$. Assume that k is a non-negative integer for which $P(k)$ is true. We will now prove that $P(k+1)$ is true as well. Assume that $n - (k+1) > \frac{2}{3}n$. We need to prove that $M_n(n-k-1) = 3(n-k-1)$, or, equivalently,

$$3(n-k-1) \geq -\frac{n}{k+1} + M_n(n-k). \quad (3.62)$$

The inequality $n-(k+1) > \frac{2}{3}n$ directly implies $n-k > \frac{2}{3}n$. The induction hypothesis $P(k)$ can be used to conclude that $M_n(n-k) = 3(n-k)$. Therefore, in order to establish that $P(k+1)$ is true (which is equivalent to (3.62)), it suffices to prove the inequality

$$3(n-k-1) \geq -\frac{n}{k+1} + 3(n-k). \quad (3.63)$$

The last inequality is equivalent to

$$\frac{n}{k+1} \geq 3. \quad (3.64)$$

Recall that we assumed that $n-(k+1) > \frac{2}{3}n$. Subtracting $\frac{2}{3}n$ from both sides gives us $\frac{1}{3}n > k+1$. Multiplying both sides by $\frac{3}{k+1}$ we get a strict inequality that clearly implies (3.64).

This completes the proof of the identity (3.62), which in turn implies $P(k+1)$. Therefore, we have completed the proof of (3.60). Thus, the optimal strategy is: If the total number of different coupons in the collection is smaller than $\frac{2}{3}n$, then pay for another coupon. Otherwise, stop playing and collect the payoff. □

3.1. DISCRETE PROBABILITY

Question 19. Let's roll the die. If the outcome is 1, 2, or 3, we stop; otherwise, if it is 4, 5, or 6, a corresponding number of dice are rolled. For example, if the first roll gives 5, then in the second round we roll 5 dice, and so on. This procedure continues for every rolled dice whose outcome is 4, 5, or 6. What is the total expected sum of all the numbers obtained at the end of the Nth round of rolls?

Answer: Let S_n be the sum of the numbers obtained in the n-th roll. We need to calculate

$$\alpha_N = \sum_{n=1}^{N} \mathbb{E}[S_n]. \tag{3.65}$$

Let D_n be the number of dice that is rolled in the n-th round. The number D_1 is equal to 1. For every $n \geq 2$, the object D_n is a random variable. For a fixed integer k, we can calculate the conditional expectation $\mathbb{E}[S_n | D_n = k]$. Once we condition on the event $\{D_n = k\}$, the expected value of S_n is the the expected sum of the rolls of k dice. Since the expected value of rolling one die is $\frac{1+2+\cdots+6}{6} = \frac{7}{2}$, it follows that

$$\mathbb{E}[S_n | D_n = k] = \frac{7}{2}k.$$

Therefore,

$$\begin{aligned}
\mathbb{E}[S_n] &= \sum_{k=0}^{\infty} \mathbb{E}[S_n | D_n = k] \cdot \mathbb{P}(D_n = k) \\
&= \sum_{k=0}^{\infty} \frac{7}{2}k \cdot \mathbb{P}(D_n = k) \\
&= \frac{7}{2} \cdot \mathbb{E}[D_n]. \tag{3.66}
\end{aligned}$$

From (3.65) and (3.66), it follows that

$$\alpha_N = \frac{7}{2} \sum_{n=1}^{N} \mathbb{E}(D_n). \tag{3.67}$$

For every $n \geq 1$ and $k \in \mathbb{N}_0$, we can calculate the conditional expectation $\mathbb{E}[D_{n+1}| D_n = k]$. When we condition on the event $\{D_n = k\}$, we are considering an experiment in which k dice are rolled. Denote by $X_1, X_2, \ldots, X_k$ the results of these k rolls and define the function $f : \{1,2,3,4,5,6\} \to \{1,2,3,4,5,6\}$ as

$$f(x) = \begin{cases} 0, & \text{if } x \leq 3, \\ x; & \text{if } x \geq 4. \end{cases}$$

The conditional expectation $\mathbb{E}[D_{n+1}| D_n = k]$ satisfies

$$\begin{aligned} \mathbb{E}[D_{n+1}| D_n = k] &= \mathbb{E}[f(X_1) + \cdots + f(X_k)] \\ &= k\mathbb{E}[f(X_1)] \\ &= k \cdot \frac{1}{6} \cdot (4+5+6) \\ &= \frac{5k}{2}. \end{aligned}$$

Then, the expectation of D_{n+1} is given by

$$\begin{aligned} \mathbb{E}[D_{n+1}] &= \sum_{k=0}^{\infty} \mathbb{E}[D_{n+1}| D_n = k] \cdot \mathbb{P}(D_n = k) \\ &= \sum_{k=0}^{\infty} \frac{5k}{2} \cdot \mathbb{P}(D_n = k) \\ &= \frac{5}{2} \mathbb{E}[D_n]. \end{aligned} \tag{3.68}$$

Since $D_1 = 1$, the recursive formula (3.68) implies that

$$\mathbb{E}[D_n] = \left(\frac{5}{2}\right)^{n-1}, \quad \forall\, n \geq 1. \tag{3.69}$$

3.1. DISCRETE PROBABILITY

From (3.67) and (3.69), we conclude that the value of α_N, which is the total expected sum of all the numbers obtained at the end of the Nth round of rolls, is

$$\begin{aligned}\alpha_N &= \frac{7}{2}\sum_{n=1}^{N}\left(\frac{5}{2}\right)^{n-1} \\ &= \frac{7}{2}\cdot\frac{\left(\frac{5}{2}\right)^{N}-1}{\frac{5}{2}-1} \\ &= \frac{7}{3}\cdot\left(\left(\frac{5}{2}\right)^{N}-1\right). \quad \square\end{aligned}$$

Question 20. We play the coin tossing game in which if tosses match, I get both coins; if they differ, you get both. You have m coins, I have n. What is the expected length of the game, that is, the number of tosses until one of us goes bankrupt?

Answer: Let us call the players A and B. Player A starts with m coins and player B starts with n coins. After each coin toss, one of the players wins a coin and the other player loses a coin. For each non-negative integer k, denote by X_k the total number of coins won by player A after k tosses. The total wealth of player A is $m + X_k$, while the total wealth of the player B is $n - X_k$.

Let T be the length of the game (the number of tosses until one of the players goes bankrupt). The random variable T is the hitting time of the set $\{-m, n\}$ by the simple symmetric random walk $(X_k)_{k=0}^{\infty}$.

We will use the following result that is proved as a part of the solution to Question 5 from Section 2.2 Random Walks and Martingales of the book "Probability and Stochastic Calculus Quant Interview Questions" [4]:

If a and b are two integers such that $a < 0 < b$, and T a time it takes for a simple symmetric random walk to reach the set $\{a, b\}$, then

$$\mathbb{E}[T] = -ab. \tag{3.70}$$

From (3.70), it follows that $E[T] = mn$. We conclude that the expected number of tosses until one of the player goes bankrupt is mn. □

3.2 Continuous Probability

Question 1. Calculate

$$\int_{-\infty}^{+\infty} (2\pi)^{-\frac{1}{2}} e^{-\frac{t^2}{2}} \Phi(t+1)\, dt,$$

where $\Phi(\cdot)$ denotes the cumulative distribution function of a standard normal random variable.

Answer: Let X and Y be independent standard normal random variables. Recalling that $\frac{1}{\sqrt{2\pi}} e^{-\frac{t^2}{2}}$ can be thought of as the probability density function $f_X(t)$ of X, we conclude that the integrand can be rewritten as

$$(2\pi)^{-\frac{1}{2}} e^{-\frac{t^2}{2}} \Phi(t+1) \;=\; f_X(t) \cdot \mathbb{P}\left(Y \leq t+1\right).$$

Thus,

$$\begin{aligned}
&\int_{-\infty}^{+\infty} (2\pi)^{-\frac{1}{2}} e^{-\frac{t^2}{2}} \Phi(t+1)\, dt \\
&= \int_{-\infty}^{+\infty} \mathbb{P}\left(Y \leq t+1\right) f_X(t)\, dt \\
&= \int_{-\infty}^{+\infty} \mathbb{P}\left(Y \leq X+1 \mid X=t\right) f_X(t)\, dt \\
&= \mathbb{P}\left(Y \leq X+1\right) = \mathbb{P}\left(Y - X \leq 1\right).
\end{aligned}$$

Since X and Y are independent and $X, Y \sim N(0,1)$, it follows that $Y - X \sim N(0,2)$. Thus, $\frac{Y-X}{\sqrt{2}}$ is a standard normal distribution and therefore

$$\mathbb{P}\left(Y - X \leq 1\right) \;=\; \mathbb{P}\left(\frac{Y-X}{\sqrt{2}} \leq \frac{1}{\sqrt{2}}\right) \;=\; \Phi\left(\frac{1}{\sqrt{2}}\right). \quad \square$$

Question 2. Suppose that $X_1, X_2, \ldots, X_n$ are independent, identically distributed random variables. Show that

$$\frac{X_1^2}{X_1^2 + X_2^2 + \ldots + X_n^2}, \quad \frac{X_2^2}{X_1^2 + X_2^2 + \ldots + X_n^2}$$

are negatively correlated.

Answer: Let

$$Y_i = \frac{X_i^2}{X_1^2 + X_2^2 + \ldots + X_n^2}, \quad \text{for } i = 1:n.$$

Note that $\sum_{i=1}^n Y_i = 1$ and therefore

$$\text{cov}\left(Y_1, \sum_{i=1}^n Y_i\right) = \text{cov}(Y_1, 1) = 0. \quad (3.71)$$

Moreover, from the bilinearity property of the covariance, we obtain that

$$\text{cov}\left(Y_1, \sum_{i=1}^n Y_i\right) = \text{var}(Y_1) + \sum_{i=2}^n \text{cov}(Y_1, Y_i). \quad (3.72)$$

From (3.71) and (3.72), it follows that

$$\text{var}(Y_1) + \sum_{i=2}^n \text{cov}(Y_1, Y_i) = 0. \quad (3.73)$$

Note that

$$\text{cov}(Y_1, Y_i) = \text{cov}(Y_1, Y_2), \quad \forall \, 2 \leq i \leq n, \quad (3.74)$$

since X_1, X_2, ..., X_n are independent identically distributed random variables.

From (3.73) and (3.74), we obtain that

$$\text{var}(Y_1) + (n-1)\text{cov}(Y_1, Y_2) = 0,$$

and therefore

$$\text{cov}(Y_1, Y_2) = -\frac{\text{var}(Y_1)}{n-1} < 0.$$

Thus, the random variables

$$\frac{X_1^2}{X_1^2 + X_2^2 + \ldots + X_n^2} = Y_1$$

3.2. CONTINUOUS PROBABILITY

and
$$\frac{X_2^2}{X_1^2 + X_2^2 + \ldots + X_n^2} = Y_2$$
are negatively correlated. □

Question 3. Let X, Y, Z be independent standard normal random variables. Find the distribution of the random variable
$$W = \frac{X + YZ}{\sqrt{1 + Z^2}}.$$

Answer: First, we find the conditional distribution of W given Z. Note that
$$X + YZ|_{Z=z} \sim X + zY \sim N\left(0, 1 + z^2\right),$$
since $X \sim N(0, 1)$ and $zY \sim N(0, z^2)$ are independent. This implies that the distribution of W, conditional on $Z = z$, is
$$W|_{Z=z} \sim \frac{X + zY}{\sqrt{1 + z^2}} \sim N(0, 1). \qquad (3.75)$$

Finally, note from (3.75) that the conditional distribution of W given Z does not depend on Z! Thus, W is independent of Z and its marginal distribution is the same as its conditional distribution given Z. Therefore, the random variable W has a standard normal distribution. □

Question 4. Calculate
$$\int\int \ldots \int e^{-\sum_{1 \leq i \leq j \leq n} x_i x_j} dx_1 dx_2 \ldots dx_n.$$

Answer: Note that
$$\sum_{1 \leq i \leq j \leq n} x_i x_j = \mathbf{x}^t \mathbf{M} \mathbf{x}, \qquad (3.76)$$

where $\mathbf{x}$ is the $n \times 1$ vector $(x_1, x_2, \ldots, x_n)^t$ and

$$\mathbf{M} = (m_{ij})_{1 \leq i \leq j \leq n}$$

is the $n \times n$ symmetric matrix with $m_{ij} = \frac{1}{2}$ for $i \neq j$ and $m_{ii} = 1$.

Next, we diagonalize $\mathbf{M}$. Since

$$\mathbf{M} = \frac{1}{2}(\mathbf{I} + \mathbf{J}),$$

where $\mathbf{I}$ is the $n \times n$ identity matrix and $\mathbf{J}$ is the all-ones $n \times n$ matrix, it is clear that the eigenvalues of $\mathbf{M}$ are $\frac{n+1}{2}$ with multiplicity 1 and $\frac{1}{2}$ with multiplicity $n - 1$. Hence, there is an orthogonal matrix $\mathbf{Q}$ such that

$$\mathbf{Q}^t \mathbf{M} \mathbf{Q} = \mathbf{Q}^{-1} \mathbf{M} \mathbf{Q} = \mathbf{D}, \qquad (3.77)$$

where $\mathbf{D} = (d_{ij})_{1 \leq i \leq j \leq n}$ is a diagonal matrix with $d_{ij} = 0$ for $i \neq j$, $d_{ii} = \frac{1}{2}$ for $1 \leq i \leq n-1$, and $d_{nn} = \frac{n+1}{2}$.

Let $\mathbf{y}$ be the $n \times 1$ vector $(y_1, y_2, \ldots, y_n)^t$ defined by the substitution $\mathbf{x} = \mathbf{Q}\mathbf{y} \iff y = \mathbf{Q}^t x$. Using (3.76), $\det(\mathbf{Q}) = 1$, and (3.77), we obtain

$$\int\int \ldots \int e^{-\sum_{1 \leq i \leq j \leq n} x_i x_j} dx_1 \, dx_2 \ldots dx_n$$
$$= \int\int \ldots \int e^{-\mathbf{x}^t \mathbf{M} \mathbf{x}} dx_1 \, dx_2 \ldots dx_n$$
$$= \int\int \ldots \int e^{-\mathbf{y}^t \mathbf{Q}^t \mathbf{M} \mathbf{Q} \mathbf{y}} \cdot \frac{1}{|\det(Q)|} dy_1 \, dy_2 \ldots dy_n$$
$$= \int\int \ldots \int e^{-\mathbf{y}^t \mathbf{D} \mathbf{y}} dy_1 \, dy_2 \ldots dy_n.$$

By expanding the expression in the exponent of the last

3.2. CONTINUOUS PROBABILITY

integral, we obtain

$$\int\int\ldots\int e^{-\sum_{1\leq i\leq j\leq n} x_i x_j}\,dx_1\,dx_2\,\ldots\,dx_n$$
$$= \int\int\ldots\int e^{-\left(\frac{y_1^2}{2}+\ldots+\frac{y_{n-1}^2}{2}+\frac{(n+1)y_n^2}{2}\right)}\,dy_1\,\ldots\,dy_n$$
$$= \int e^{-\frac{y_1^2}{2}}\,dy_1 \int e^{-\frac{y_2^2}{2}}\,dy_2\ldots \int e^{-\frac{y_{n-1}^2}{2}}\,dy_{n-1} \times$$
$$\times \int e^{-\frac{(n+1)y_n^2}{2}}\,dy_n$$
$$= \left(\sqrt{2\pi}\right)^{n-1}\cdot\sqrt{\frac{2\pi}{n+1}} = \frac{\left(\sqrt{2\pi}\right)^n}{\sqrt{n+1}}.\quad\square$$

Question 5. Let $A = (a_{i,j})$ and $B = (b_{i,j})$ be two symmetric non-negative semidefinite $n \times n$ matrices. Let $C = (c_{i,j})$ be the $n \times n$ matrix given by $c_{i,j} = a_{i,j}b_{i,j}$. Show that C is also symmetric non–negative semidefinite.

Answer: Recall that every covariance matrix must be non-negative semidefinite. The converse is also true, that is, every symmetric non–negative definite matrix is a covariance matrix of some multivariate distribution! Indeed, every non–negative semidefinite $n \times n$ matrix M can be diagonalized as $QD^2Q^\top$, where Q is an orthogonal matrix and D is a nonnegative diagonal matrix. Letting Z denote a random vector with the standard multivariate normal distribution $N(0_{n\times 1}, I_{n\times n})$, it is clear that M is the covariance matrix of

$$V = QDZ.$$

Note that in this case $V \sim N(0_{n\times 1}, M)$.

Hence, using this observation, we can assume that the symmetric non–negative definite $n\times n$ matrices $A = (a_{i,j})$

and $B = (b_{i,j})$ are covariance matrices of independent multivariate normal random vectors

$$\begin{aligned} X &= (X_1, X_2, \ldots, X_n)^t \sim N(0_{n \times 1}, A); \\ Y &= (Y_1, Y_2, \ldots, Y_n)^t \sim N(0_{n \times 1}, B), \end{aligned}$$

respectively. In particular,

$$\begin{aligned} a_{i,j} &= \text{cov}(X_i, X_j) \\ &= \mathbb{E}[X_i X_j] - \mathbb{E}[X_i]\mathbb{E}[X_j] \\ &= \mathbb{E}[X_i X_j]; \quad (3.78) \\ b_{i,j} &= \text{cov}(Y_i, Y_j) \\ &= \mathbb{E}[Y_i Y_j] - \mathbb{E}[Y_i]\mathbb{E}[Y_j] \\ &= \mathbb{E}[Y_i Y_j]. \quad (3.79) \end{aligned}$$

Then, the covariance matrix of the random vector $W = (X_1Y_1, X_2Y_2, \ldots, X_nY_n)^t$ is exactly the matrix $C = (c_{i,j})$ given by

$$\begin{aligned} c_{i,j} &= \text{cov}(W_i, W_j) \\ &= \mathbb{E}[X_i Y_i X_j Y_j] - \mathbb{E}[X_i Y_i]\mathbb{E}[X_j Y_j] \\ &= \mathbb{E}[X_i X_j Y_i Y_j] - \mathbb{E}[X_i]\mathbb{E}[Y_i]\mathbb{E}[X_j]\mathbb{E}[Y_j] \\ &= \mathbb{E}[X_i X_j]\mathbb{E}[Y_i Y_j] \\ &= a_{i,j} b_{i,j}, \end{aligned}$$

where, for the last equality, we used (3.78) and (3.79).

Being a covariance matrix, this shows that the matrix C is also symmetric non–negative semidefinite. $\square$

Question 6. Find

$$\lim_{n \to \infty} \sum_{k=n}^{3n} \binom{k-1}{n-1} \left(\frac{1}{3}\right)^n \left(\frac{2}{3}\right)^{k-n}.$$

3.2. CONTINUOUS PROBABILITY

Answer: Let

$$L = \lim_{n\to\infty} \sum_{k=n}^{3n} \binom{k-1}{n-1} \left(\frac{1}{3}\right)^n \left(\frac{2}{3}\right)^{k-n}.$$

Imagine that you repeatedly toss a biased coin with the probability of tossing heads in a single toss equal to $p = \frac{1}{3}$. Denote by A_n the event that the nth head appears among the first $3n$ tosses. Then,

$$A_n = \bigcup_{k=n}^{3n} A_n^k,$$

where A_n^k denotes the event that the nth head appears exactly on the kth toss, for $n \leq k \leq 3n$. Since the first $n-1$ heads must have appeared among the first $k-1$ tosses, we find that

$$\begin{aligned}\mathbb{P}\left(A_n^k\right) &= p \cdot \binom{k-1}{n-1} p^{n-1} \cdot (1-p)^{k-n} \\ &= \binom{k-1}{n-1} \left(\frac{1}{3}\right)^n \left(\frac{2}{3}\right)^{k-n}.\end{aligned}$$

Then,

$$L = \lim_{n\to\infty} \sum_{k=n}^{3n} \mathbb{P}\left(A_n^k\right) = \lim_{n\to\infty} \mathbb{P}(A_n).$$

From a different perspective, A_n is equivalent to the event

$$\sum_{k=1}^{3n} X_k \geq n,$$

where the random variable X_k, for $1 \leq k \leq n$, denotes the number of heads in the kth Note that X_k's are independent Bernoulli random variables with parameter $p = \frac{1}{3}$.

Their expected values are $\mathbb{E}[X_k] = \frac{1}{3}$. The variances are $\text{var}(X_k) = p(1-p) = \frac{2}{9}$. From the Central Limit Theorem, we deduce that, as $n \to \infty$, $\frac{\sum_{k=1}^{3n}(X_k - \frac{1}{3})}{\sqrt{\frac{2}{9} \cdot 3n}}$ converges in distribution to a standard normal random variable Z. Therefore,

$$\begin{aligned} L &= \lim_{n \to \infty} \mathbb{P}(A_n) \\ &= \lim_{n \to \infty} \mathbb{P}\left(\sum_{k=1}^{3n} X_k \geq n\right) \\ &= \lim_{n \to \infty} \mathbb{P}\left(\frac{\sum_{k=1}^{n}(X_k - \frac{1}{3})}{\sqrt{\frac{2}{9} \cdot 3n}} \geq 0\right) \\ &= \mathbb{P}(Z \geq 0). \end{aligned}$$

Since Z has a normal distribution with mean 0, we obtain that $L = \frac{1}{2}$. $\square$

Question 7. Let $X_1, X_2, \ldots, X_n$ be independent identically distributed uniform random variables on $[-1, 1]$. Define the random variable

$$X = \frac{\sum_{i=1}^{n} X_i}{\sqrt{\sum_{i=1}^{n} X_i^2}}.$$

Approximate the probability of $X > 1$, as $n \to \infty$.

Answer: Since $X_1, X_2, \ldots, X_n$ are uniform random variables on $[-1, 1]$, it follows that, for every $i = 1 : n$,

$$\mathbb{E}[X_i] = 0 \text{ and } \text{Var}(X) = \mathbb{E}[X_i^2] = \frac{1}{3}.$$

Note that the random variable X can be written as

$$X = \frac{\sqrt{3} \cdot \frac{1}{\sqrt{n}} \sum_{i=1}^{n} X_i}{\sqrt{3} \cdot \sqrt{\frac{1}{n} \sum_{i=1}^{n} X_i^2}}. \qquad (3.80)$$

3.2. CONTINUOUS PROBABILITY

Using the Weak Law of Large Numbers and the continuous mapping theorem, we obtain the following convergence in probability as $n \to \infty$:

$$\sqrt{\frac{1}{n}\sum_{i=1}^{n} X_i^2} \xrightarrow{p} \sqrt{\mathbb{E}[X_i^2]} = \frac{1}{\sqrt{3}}. \qquad (3.81)$$

From the Central Limit Theorem, we also have the convergence in distribution

$$\frac{\frac{1}{n}\sum_{i=1}^{n} X_i}{\frac{1}{\sqrt{3}}\frac{1}{\sqrt{n}}} = \frac{\sqrt{3}}{\sqrt{n}}\sum_{i=1}^{n} X_i \xrightarrow{D} Z, \qquad (3.82)$$

as $n \to \infty$, where $Z \sim N(0,1)$ is the standard normal random variable.

By combining (3.81), (3.82) with (3.80) and using Slutsky's theorem, we conclude that

$$X \xrightarrow{D} \frac{Z}{\sqrt{3}/\sqrt{3}} = Z,$$

as $n \to \infty$.

Therefore, as $n \to \infty$, we can approximate $\mathbb{P}(X > 1)$ with $\mathbb{P}(Z > 1) = 1 - \Phi(1)$, where $\Phi(\cdot)$ denotes the cumulative density function of the standard normal distribution. $\square$

Question 8. Calculate

$$\lim_{n\to\infty} \int_{[0,1]^n} \frac{x_1^5 + \ldots + x_n^5}{x_1^4 + \ldots + x_n^4}\, dx_1 \ldots dx_n.$$

Answer: Let $X_1, X_2, \ldots,$ be a sequence of independent random variables with uniform distribution on the interval $[0, 1]$. Note that

$$E\left[\frac{X_1^5 + \cdots + X_n^5}{X_1^4 + \cdots + X_n^4}\right] = \int_{[0,1]^n} \frac{x_1^5 + \ldots + x_n^5}{x_1^4 + \ldots + x_n^4}\, dx_1 \ldots dx_n.$$

Thus, we need to evaluate

$$L = \lim_{n \to \infty} \mathbb{E}\left[\frac{X_1^5 + \cdots + X_n^5}{X_1^4 + \cdots + X_n^4}\right]. \quad (3.83)$$

We will use Central Limit Theorem to prove that

$$L = \frac{\mu_5}{\mu_4},$$

where $\mu_5 = \frac{1}{6}$ and $\mu_4 = \frac{1}{5}$ are expected values of the fifth and fourth moment of the uniform $[0, 1]$ random variable.

If U is a uniform random variable on interval $[0, 1]$, its k-th moment satisfies

$$\mu_k = \mathbb{E}\left[U^k\right] = \int_0^1 u^k \, du = \frac{1}{k+1}, \quad \forall \, k \geq 1.$$

The standard deviations σ_4 and σ_5 of the random variables U^4 and U^5 are

$$\sigma_4 = \frac{4}{15} \quad \text{and} \quad \sigma_5 = \frac{5}{6\sqrt{11}}. \quad (3.84)$$

Let us fix a real number $\varepsilon > 0$. There exists a positive real number ω such that

$$\frac{1}{\sqrt{2\pi}} \int_{-\omega}^{\omega} e^{-\frac{u^2}{2}} \, du > 1 - \frac{\varepsilon}{8}. \quad (3.85)$$

The numbers $\varepsilon > 0$ and $\omega > 0$ are now fixed. According to the Central Limit Theorem, there exists an integer n_0 such that, for every $n \geq n_0$, the following relations hold:

$$\mathbb{P}\left(\frac{X_1^4 + \cdots + X_n^4 - n\mu_4}{\sqrt{n} \cdot \sigma_4} \in (-\omega, \omega)\right) \quad (3.86)$$

$$-\frac{1}{\sqrt{2\pi}} \int_{-\omega}^{\omega} e^{-\frac{u^2}{2}} \, du \in \left(-\frac{\varepsilon}{8}, \frac{\varepsilon}{8}\right)$$

3.2. CONTINUOUS PROBABILITY

and

$$\mathbb{P}\left(\frac{X_1^5 + \cdots + X_n^5 - n\mu_5}{\sqrt{n}\cdot\sigma_5} \in (-\omega,\omega)\right) \qquad (3.87)$$
$$-\frac{1}{\sqrt{2\pi}}\int_{-\omega}^{\omega} e^{-\frac{u^2}{2}}\,du \ \in \ \left(-\frac{\varepsilon}{8},\frac{\varepsilon}{8}\right).$$

Denote by B_n^4 and B_n^5 the following events

$$B_n^4 \ = \ \left\{\frac{X_1^4+\cdots+X_n^4 - n\mu_4}{\sqrt{n}\cdot\sigma_4} \notin (-\omega,\omega)\right\} \quad (3.88)$$
$$B_n^5 \ = \ \left\{\frac{X_1^5+\cdots+X_n^5 - n\mu_5}{\sqrt{n}\cdot\sigma_5} \notin (-\omega,\omega)\right\}. \quad (3.89)$$

From (3.85) and (3.86) we conclude that $\mathbb{P}\left(B_n^4\right) \leq \frac{\varepsilon}{4}$. Similarly, we obtain that the probability of the event B_n^5 is at most $\frac{\varepsilon}{4}$. Let $G_n = \left(B_n^4 \cup B_n^5\right)^C$. The probability of the event G_n is at least $1 - \frac{\varepsilon}{2}$. Since X_i is a uniform random variable on $[0,1]$, we must have $X_i \leq 1$ and $X_i^5 \leq X_i^4$. Therefore, the random variable

$$Z_n \ = \ \frac{X_1^5+\cdots+X_n^5}{X_1^4+\cdots+X_n^4}$$

must take values in $[0,1]$. Hence, for $n \geq n_0$ we have

$$\mathbb{E}\left[Z_n\right] \ = \ \mathbb{E}[Z_n \cdot 1_{G_n}] + \mathbb{E}\left[Z_n \cdot 1_{G_n^C}\right]. \quad (3.90)$$

The probability of G_n^C is at most $\varepsilon/2$ and $Z_n \in [0,1]$. Therefore,

$$\mathbb{E}\left[Z_n \cdot 1_{G_n^C}\right] \ \in \ \left(0,\frac{\varepsilon}{2}\right). \quad (3.91)$$

Let us denote the numerator and denominator of Z_n by S_n^5 and S_n^4. More precisely,

$$S_n^5 \ = \ X_1^5 + \cdots + X_n^5$$
$$S_n^4 \ = \ X_1^4 + \cdots + X_n^4.$$

On the event G_n we have

$$S_n^5 \in \left(n\mu_5 - \sqrt{n} \cdot \omega \cdot \sigma_5, n\mu_5 + \sqrt{n} \cdot \omega \cdot \sigma_5\right),$$
$$S_n^4 \in \left(n\mu_4 - \sqrt{n} \cdot \omega \cdot \sigma_4, n\mu_4 + \sqrt{n} \cdot \omega \cdot \sigma_4\right).$$

From the last two relations we conclude that on G_n the following two inequalities hold:

$$Z_n = \frac{S_n^5}{S_n^4} \leq \frac{n\mu_5 + \sqrt{n} \cdot \omega \cdot \sigma_5}{n\mu_4 - \sqrt{n} \cdot \omega \cdot \sigma_4} = \frac{\frac{\mu_5}{\mu_4} + \frac{\omega \cdot \sigma_5}{\sqrt{n} \cdot \mu_4}}{1 - \frac{\omega \cdot \sigma_4}{\sqrt{n} \cdot \mu_4}}, \quad (3.92)$$

$$Z_n = \frac{S_n^5}{S_n^4} \geq \frac{n\mu_5 - \sqrt{n} \cdot \omega \cdot \sigma_5}{n\mu_4 + \sqrt{n} \cdot \omega \cdot \sigma_4} = \frac{\frac{\mu_5}{\mu_4} - \frac{\omega \cdot \sigma_5}{\sqrt{n} \cdot \mu_4}}{1 + \frac{\omega \cdot \sigma_4}{\sqrt{n} \cdot \mu_4}}. \quad (3.93)$$

We now use (3.90), (3.91), (3.92), and (3.93) to obtain

$$\mathbb{E}\left[Z_n\right] \leq \frac{\frac{\mu_5}{\mu_4} + \frac{\omega \cdot \sigma_5}{\sqrt{n} \cdot \mu_4}}{1 - \frac{\omega \cdot \sigma_4}{\sqrt{n} \cdot \mu_4}} + \frac{\varepsilon}{2}, \text{ and}$$

$$\mathbb{E}\left[Z_n\right] \geq \frac{\frac{\mu_5}{\mu_4} - \frac{\omega \cdot \sigma_5}{\sqrt{n} \cdot \mu_4}}{1 + \frac{\omega \cdot \sigma_4}{\sqrt{n} \cdot \mu_4}}.$$

Taking $n \to \infty$ in the last two inequalities gives us

$$\frac{\mu_5}{\mu_4} \leq \liminf_{n \to \infty} \mathbb{E}\left[Z_n\right] \leq \limsup_{n \to \infty} \mathbb{E}\left[Z_n\right] \leq \frac{\mu_5}{\mu_4} + \frac{\varepsilon}{2}.$$

Since the previous inequality holds for every $\varepsilon > 0$, we obtain that the required limit is $\frac{\mu_5}{\mu_4} = \frac{5}{6}$. $\square$

Question 9. Two real numbers X and Y are chosen at random in the interval $(0, 1)$ with respect to the uniform distribution. What is the probability that the closest integer to X/Y is even? Express your answer in terms of π.

Answer: The point (X, Y) is distributed uniformly at random inside the unit square $[0, 1] \times [0, 1]$. Denote by N

3.2. CONTINUOUS PROBABILITY

the integer closest to $\frac{X}{Y}$. We need to compute

$$p = \sum_{k=0}^{\infty} \mathbb{P}(N = 2k). \qquad (3.94)$$

Note that

$$N = 0 \iff 0 \le \frac{X}{Y} \le \frac{1}{2} \iff Y \ge 2X.$$

Thus, the event $N = 0$ is the set of all points (X, Y) inside the unit square $[0, 1] \times [0, 1]$ that satisfy $Y \ge 2X$. This region is a triangle with height 1 and base $\frac{1}{2}$ and therefore

$$\mathbb{P}(N = 0) = \frac{1}{4}. \qquad (3.95)$$

Let $k \ge 1$. Note that

$$N = 2k \iff 2k - \frac{1}{2} \le \frac{X}{Y} \le 2k + \frac{1}{2}$$
$$\iff \frac{2}{4k+1}X \le Y \le \frac{2}{4k-1}X.$$

Thus, the event $\{N = 2k\}$ is the set of all points (X, Y) inside the unit square $[0, 1] \times [0, 1]$ that satisfy

$$\frac{2}{4k+1}X \le Y \le \frac{2}{4k-1}X.$$

This region is a triangle with height 1 and base

$$\frac{2}{4k-1} - \frac{2}{4k+1}.$$

We conclude that

$$\mathbb{P}(N = 2k) = \frac{1}{4k-1} - \frac{1}{4k+1}, \quad \forall\, k \ge 1. \qquad (3.96)$$

From (3.94), (3.95), and (3.96), we obtain that

$$p = \frac{1}{4} + \sum_{k=1}^{\infty} \left(\frac{1}{4k-1} - \frac{1}{4k+1} \right). \qquad (3.97)$$

We now have to express (3.97) in terms of π. Recall the Leibniz formula

$$\begin{aligned}
\frac{\pi}{4} &= \arctan(x)\Big|_0^1 = \int_0^1 \frac{1}{1+x^2}\, dx \\
&= \int_0^1 \frac{1}{1-(-x^2)}\, dx \\
&= \int_0^1 \sum_{k=0}^{\infty} (-1)^k x^{2k}\, dx \\
&= \sum_{k=0}^{\infty} (-1)^k \int_0^1 x^{2k}\, dx \\
&= \sum_{k=0}^{\infty} (-1)^k \frac{1}{2k+1} \quad (3.98)
\end{aligned}$$

Note that

$$\sum_{k=0}^{\infty} (-1)^k \frac{1}{2k+1} = 1 - \sum_{k=1}^{\infty}\left(\frac{1}{4k-1} - \frac{1}{4k+1}\right). \quad (3.99)$$

From (3.97), (3.98), and (3.99), we obtain that

$$p = \frac{1}{4} + \left(1 - \frac{\pi}{4}\right) = \frac{5-\pi}{4}.$$

We conclude that the probability that the closest integer to X/Y is even equals $\frac{5-\pi}{4}$. $\square$

Question 10. A random vector $X = (X_1, X_2, \ldots, X_n)$ is distributed uniformly at random on the unit sphere in $\mathbb{R}^n$. Find $\mathbb{E}\left[X_1^2\right]$ and $\mathbb{E}\left[X_1^4\right]$.

Answer: Since

$$\sum_{i=1}^{n} X_i^2 = 1, \quad (3.100)$$

we obtain that

$$1 = \mathbb{E}\left[\sum_{i=1}^{n} X_i^2\right] = \sum_{i=1}^{n} \mathbb{E}\left[X_i^2\right] = n\mathbb{E}\left[X_1^2\right],$$

where the last step follows from the fact that $X_1, X_2, \ldots, X_n$ are identically distributed. Therefore,

$$\mathbb{E}\left[X_1^2\right] = \frac{1}{n}.$$

Moreover,

$$\left(\sum_{i=1}^{n} X_i^2\right)^2 = 1,$$

see (3.100), and therefore the linearity of expectation implies that

$$\begin{aligned}
1 &= \mathbb{E}\left[\left(\sum_{i=1}^{n} X_i^2\right)^2\right] \\
&= \sum_{i=1}^{n} \mathbb{E}\left[X_i^4\right] + \sum_{i \neq j} \mathbb{E}\left[X_i^2 X_j^2\right] \\
&= n\mathbb{E}\left[X_1^4\right] + n(n-1)\mathbb{E}\left[X_1^2 X_2^2\right]. \quad (3.101)
\end{aligned}$$

In the last step we again used that $X_1, X_2, \ldots, X_n$ are identically distributed.

Next, we use the fact that the vectors

$$(X_1, X_2) \quad \text{and} \quad \left((X_1 + X_2)/\sqrt{2}, (X_1 - X_2)/\sqrt{2}\right)$$

are identically distributed.

Indeed, the joint densities $f_{(X_1+X_2)/\sqrt{2},(X_1-X_2)/\sqrt{2}}$ and f_{X_1,X_2} are related via the Jacobian as follows

$$f_{(X_1+X_2)/\sqrt{2},(X_1-X_2)/\sqrt{2}} = \frac{f_{X_1,X_2}}{\left|\det\left[\frac{\partial\left(\frac{X_1+X_2}{\sqrt{2}}, \frac{X_1-X_2}{\sqrt{2}}\right)}{\partial(X_1,X_2)}\right]\right|}. \quad (3.102)$$

Here,

$$\det\left[\frac{\partial\left(\frac{X_1+X_2}{\sqrt{2}}, \frac{X_1-X_2}{\sqrt{2}}\right)}{\partial(X_1, X_2)}\right]$$
$$= \det\begin{bmatrix} \frac{\partial((X_1+X_2)/\sqrt{2})}{\partial X_1} & \frac{\partial((X_1+X_2)/\sqrt{2})}{\partial X_2} \\ \frac{\partial((X_1-X_2)/\sqrt{2})}{\partial X_1} & \frac{\partial((X_1-X_2)/\sqrt{2})}{\partial X_2} \end{bmatrix}$$
$$= \det\begin{bmatrix} 1/\sqrt{2} & 1/\sqrt{2} \\ 1/\sqrt{2} & -1/\sqrt{2} \end{bmatrix}$$
$$= -1. \tag{3.103}$$

From (3.102) and (3.103), we conclude that

$$f_{\frac{X_1+X_2}{\sqrt{2}}, \frac{X_1-X_2}{\sqrt{2}}} = f_{X_1, X_2},$$

as claimed. In particular, we have that

$$\mathbb{E}\left[\left(\frac{X_1+X_2}{\sqrt{2}}\right)^2 \left(\frac{X_1-X_2}{\sqrt{2}}\right)^2\right] = \mathbb{E}\left[X_1^2 X_2^2\right]. \tag{3.104}$$

By expanding the left–hand side of (3.104), we obtain that

$$\mathbb{E}\left[\left(\frac{X_1+X_2}{\sqrt{2}}\right)^2 \left(\frac{X_1-X_2}{\sqrt{2}}\right)^2\right]$$
$$= \frac{1}{4}\mathbb{E}\left[\left(X_1^2 - X_2^2\right)^2\right]$$
$$= \frac{1}{4}\left(\mathbb{E}\left[X_1^4\right] - 2\mathbb{E}\left[X_1^2 X_2^2\right] + \mathbb{E}\left[X_2^4\right]\right)$$
$$= \frac{1}{2}\mathbb{E}\left[X_1^4\right] - \frac{1}{2}\mathbb{E}\left[X_1^2 X_2^2\right]. \tag{3.105}$$

Then, from (3.104) and (3.105), we find that

$$\mathbb{E}\left[X_1^4\right] = 3\mathbb{E}\left[X_1^2 X_2^2\right]. \tag{3.106}$$

3.2. CONTINUOUS PROBABILITY

From (3.101) and (3.106), we obtain that

$$\begin{aligned}1 &= 3n\,\mathbb{E}\left[X_1^2 X_2^2\right] + n(n-1)\,\mathbb{E}\left[X_1^2 X_2^2\right] \\ &= (n^2 + 2n)\,\mathbb{E}\left[X_1^2 X_2^2\right],\end{aligned}$$

and therefore

$$\mathbb{E}\left[X_1^2 X_2^2\right] = \frac{1}{n(n+2)}. \tag{3.107}$$

From (3.106) and (3.107), we conclude that

$$\mathbb{E}\left[X_1^4\right] = \frac{3}{n(n+2)}. \quad \square$$

Question 11. The number of people that have arrived at a train station by time t is a Poisson random variable with rate λt. If the train arrives at the station at a time uniformly distributed over $(0, T)$ (and independent of when the passengers arrive), what are the expected value and the variance of the number of passengers that enter this train?

Answer: Denote by Z the number of people that enter the train and by A the arrival time of the train. We know that, conditioned on the event $\{A = t\}$, the random variable Z has Poisson distribution with parameter λt. Therefore, the conditional expectations of Z and Z^2 are

$$\begin{aligned}\mathbb{E}\left[Z | A = t\right] &= \lambda t \\ \mathbb{E}\left[Z^2 | A = t\right] &= \lambda t + \lambda^2 t^2.\end{aligned}$$

The expectation of Z is

$$\begin{aligned}\mathbb{E}[Z] &= \frac{1}{T} \int_0^T \mathbb{E}\left[Z | A = t\right] dt \\ &= \frac{1}{T} \int_0^T \lambda t \, dt \\ &= \frac{\lambda T}{2}. \tag{3.108}\end{aligned}$$

Similarly,

$$\begin{aligned}
\mathbb{E}\left[Z^2\right] &= \frac{1}{T}\int_0^T \mathbb{E}\left[Z^2 \mid A = t\right] dt \\
&= \frac{1}{T}\int_0^T \left(\lambda t + \lambda^2 t^2\right) dt \\
&= \frac{\lambda T}{2} + \frac{\lambda^2 T^2}{3}.
\end{aligned} \qquad (3.109)$$

From (3.108) and (3.109), we obtain that

$$\text{var}(Z) = \mathbb{E}\left[Z^2\right] - (\mathbb{E}[Z])^2 = \frac{\lambda T}{2} + \frac{\lambda^2 T^2}{12}.$$

We conclude that the expected value and the variance of the number of passengers that enter the train are $\frac{\lambda T}{2}$ and $\frac{\lambda T}{2} + \frac{\lambda^2 T^2}{12}$, respectively. □

Question 12. An infinite sheet of paper has inscribed on it a set of horizontal lines D units apart and a set of vertical lines D units apart. A needle of length L (where $L < D$) is twirled and tossed on the paper. What is the expected number of lines crossed by the needle? What is the probability that the needle crosses a line?

Answer: Since $L < D$, the needle can intersect at most one horizontal and at most one vertical line. Denote by X the distance from the midpoint of the needle to the closest horizontal line.[1] Denote by Y the distance from the midpoint of the needle to the closest vertical line. Let θ denote the angle between the vertical line through the midpoint of the needle and the line containing the needle itself, as in the figure below.

---
[1] The probability that the needle will land on the sheet of paper so that its midpoint is exactly halfway between two consecutive horizontal lines is zero, so the horizontal line closest to the midpoint of the needle is well-defined.

3.2. CONTINUOUS PROBABILITY

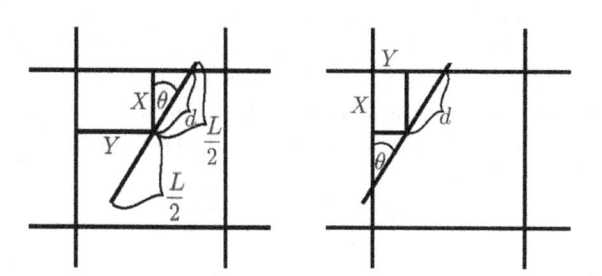

Note that the position of the needle is completely determined by the values of X, Y, and θ. The random variables X, Y, and θ are independent; X and Y are uniformly distributed on $\left(0, \frac{D}{2}\right)$, while θ is uniformly distributed on $\left(0, \frac{\pi}{2}\right)$. Denote by A the event that the needle crosses a horizontal line and by B the event that the needle crosses a vertical line.

Since the needle can intersect at most one horizontal and at most one vertical line, the expected number of lines crossed by the needle is precisely

$$\mathbb{P}(A) + \mathbb{P}(B) = 2\mathbb{P}(A).$$

The last equality follows from $\mathbb{P}(A) = \mathbb{P}(B)$, a fact easily established by rotating the sheet of paper by $\pi/2$. Similarly, the probability that the needle crosses a line is precisely

$$\mathbb{P}(A) + \mathbb{P}(B) - \mathbb{P}(A \cap B) = 2\mathbb{P}(A) - \mathbb{P}(A \cap B).$$

Hence, we need to compute $\mathbb{P}(A)$ and $\mathbb{P}(A \cap B)$. Denote by d the distance from the midpoint of the needle to the point of intersection of the line containing the needle itself with the horizontal line (see figure).

184   CHAPTER 3. PROBABILISTIC PUZZLES

Note that $d = \frac{X}{\cos\theta}$. The event A occurs if and only if $d < \frac{L}{2}$, that is, if and only if $\frac{X}{\cos\theta} < \frac{L}{2}$. Similarly, the event B occurs if and only if $\frac{Y}{\sin\theta} < \frac{L}{2}$.

Since X and θ are independent, their joint density $f_{X,\theta}$ is the product of marginal densities and, thus, satisfies

$$f_{X,\theta}(u,v) = \frac{1}{D/2} \cdot \frac{1}{\pi/2}.$$

By integrating the joint density over an appropriate region, we obtain that

$$\begin{aligned}
\mathbb{P}(A) &= \mathbb{P}\left(\frac{X}{\cos\theta} < \frac{L}{2}\right) = \mathbb{P}\left(X < \frac{L}{2}\cos\theta\right) \\
&= \int_0^{\frac{\pi}{2}} \int_0^{\frac{L}{2}\cos\theta} \frac{1}{D/2} \cdot \frac{1}{\pi/2}\, dx\, d\theta \\
&= \frac{4}{\pi D} \int_0^{\frac{\pi}{2}} \frac{L}{2} \cos\theta\, d\theta \\
&= \frac{2L}{\pi D} \sin\theta \Big|_0^{\pi/2} \\
&= \frac{2L}{\pi D}.
\end{aligned}$$

Similarly, since X, Y, and θ are independent, their joint density is the product of marginal densities and, thus, is equal to $\left(\frac{1}{D/2}\right)^2 \cdot \frac{1}{\pi/2}$. Then,

$$\begin{aligned}
&\mathbb{P}(A \cap B) \\
&= \mathbb{P}\left(\frac{X}{\cos\theta} < \frac{L}{2} \text{ and } \frac{Y}{\sin\theta} < \frac{L}{2}\right) \\
&= \mathbb{P}\left(X < \frac{L}{2}\cos\theta \text{ and } Y < \frac{L}{2}\sin\theta\right) \\
&= \int_0^{\frac{\pi}{2}} \int_0^{\frac{L}{2}\cos\theta} \int_0^{\frac{L}{2}\sin\theta} \left(\frac{1}{D/2}\right)^2 \cdot \frac{1}{\pi/2}\, dy\, dx\, d\theta.
\end{aligned}$$

3.2. CONTINUOUS PROBABILITY

The last integral can be evaluated as follows:

$$\begin{aligned}
\mathbb{P}(A \cap B) &= \frac{8}{\pi D^2} \cdot \frac{L^2}{4} \int_0^{\frac{\pi}{2}} \sin\theta \cos\theta \, d\theta \\
&= \frac{L^2}{\pi D^2} \int_0^{\frac{\pi}{2}} \sin(2\theta) \, d\theta \\
&= \frac{L^2}{\pi D^2} \cdot \left(-\frac{1}{2} \cos(2\theta) \right) \Big|_0^{\pi/2} \\
&= \frac{L^2}{\pi D^2}.
\end{aligned}$$

We conclude that the expected number of lines crossed by the needle is

$$2\,\mathbb{P}(A) = \frac{4L}{\pi D},$$

while the probability that the needle crosses a line is

$$2\,\mathbb{P}(A) - \mathbb{P}(A \cap B) = \frac{4L}{\pi D} - \frac{L^2}{\pi D^2}. \quad \square$$

Question 13. Let $\{T_k\}$ be an independent identically distributed sample from the exponential distribution with parameter λ. Let $S_n = \sum_{k=1}^{n} T_k$ and define

$$N_t = \max\{k : S_k \leq t\}.$$

Find the distribution of N_t for $t > 0$.

Answer: We will prove that, for $t > 0$, the random variable N_t has a Poisson distribution with parameter λt.

Note that N_t is an integer-valued random variable and its values are non–negative integers. The event $\{N_t \geq m\}$ can be described as

$$\{N_t \geq m\} = \{\max\{k : S_k \leq t\} \geq m\}. \qquad (3.110)$$

Moreover, if $\max\{k : S_k \leq t\} \geq m$, then $S_m \leq t$ since $(S_n)_{n=0}^{\infty}$ is an increasing sequence. Therefore,

$$\{\max\{k : S_k \leq t\} \geq m\} \subseteq \{S_m \leq t\}. \qquad (3.111)$$

On the other hand, if m is a positive integer such that $S_m \leq t$, then m must belong to the set $\{k : S_k \leq t\}$. Since every number is smaller than the maximum of the set to which it belongs, $m \leq \max\{k : S_k \leq t\}$ and therefore

$$\{\max\{k : S_k \leq t\} \geq m\} \supseteq \{S_m \leq t\}. \qquad (3.112)$$

From (3.111) and (3.112), we find that

$$\{\max\{k : S_k \leq t\} \geq m\} = \{S_m \leq t\}. \qquad (3.113)$$

Then, (3.110) and (3.113) imply that

$$\mathbb{P}(N_t \geq m) = \mathbb{P}(S_m \leq t). \qquad (3.114)$$

Recall that the probability density function of the Gamma random variable with parameters (α, λ) is given by

$$f(x) = \frac{\lambda^{\alpha}}{\Gamma(\alpha)} \cdot e^{-\lambda x} \cdot x^{\alpha-1} \cdot 1_{(0,+\infty)}(x), \qquad (3.115)$$

where

$$\Gamma(\alpha) = \int_0^{+\infty} e^{-x} x^{\alpha-1} \, dx.$$

When α is a positive integer, the Gamma function $\Gamma(\alpha)$ has the value $\Gamma(\alpha) = (\alpha - 1)!$. An exponential random variable with parameter λ is also a Gamma random variable with parameters $(1, \lambda)$. The sum of independent Gamma random variables with parameters (α_1, γ) and (α_2, γ) is a gamma random variable with parameter $(\alpha_1 + \alpha_2, \gamma)$. Therefore, S_m has Gamma distribution with parameters (m, λ). We can now use the probability density function given by (3.115) with $\alpha = m$ to evaluate

3.2. CONTINUOUS PROBABILITY

$\mathbb{P}(S_m \leq t)$ as follows:

$$\begin{aligned} \mathbb{P}(S_m \leq t) &= \int_0^t \frac{\lambda^m}{\Gamma(m)} \cdot e^{-\lambda u} \cdot u^{m-1} \, du \\ &= \int_0^t \frac{\lambda^m}{(m-1)!} \cdot e^{-\lambda u} \cdot u^{m-1} \, du. \end{aligned} \quad (3.116)$$

We will use the substitution $v = \lambda u$ for the integral from (3.116). We find that

$$\mathbb{P}(S_m \leq t) = \int_0^{\lambda t} \frac{v^{m-1}}{(m-1)!} \cdot e^{-v} \, dv.$$

Denote the last integral by

$$I(m, \lambda, t) = \int_0^{\lambda t} \frac{v^{m-1}}{(m-1)!} \cdot e^{-v} \, dv.$$

Then,

$$\mathbb{P}(S_m \leq t) = I(m, \lambda, t). \quad (3.117)$$

For $m = 1$,

$$I(1, \lambda, t) = \int_0^{\lambda t} e^{-v} \, dv = 1 - e^{-\lambda t}. \quad (3.118)$$

If $m \geq 2$ is a positive integer, we use integration by parts repeatedly. Define $f(v) = \frac{v^{m-1}}{(m-1)!}$ and $g(v) = -e^{-v}$. Then, the integration by parts

$$\int f(v) g'(v) \, dv = f(v) g(v) - \int g(v) f'(v) \, dv$$

gives us

$$\begin{aligned} &I(m, \lambda, t) \\ &= \left. -\frac{v^{m-1}}{(m-1)!} \cdot e^{-v} \right|_{v=0}^{v=\lambda t} + \int_0^{\lambda t} \frac{v^{m-2}}{(m-2)!} e^{-v} \, dv \\ &= -\frac{(\lambda t)^{m-1}}{(m-1)!} \cdot e^{-\lambda t} + I(m-1, \lambda, t). \end{aligned} \quad (3.119)$$

Using induction and the identities (3.118) and (3.119), we obtain that

$$I(m, \lambda, t) = 1 - e^{-\lambda t} \sum_{i=1}^{m} \frac{(\lambda t)^{m-1}}{(m-1)!}. \qquad (3.120)$$

From (3.114), (3.117), and (3.120), it follows that

$$\mathbb{P}(N_t \geq m) = 1 - e^{-\lambda t} \sum_{i=1}^{m} \frac{(\lambda t)^{i-1}}{(i-1)!}. \qquad (3.121)$$

By using (3.121) with m and $m+1$, we conclude that

$$\begin{aligned} \mathbb{P}(N_t = m) &= \mathbb{P}(N_t \geq m) - \mathbb{P}(N_t \geq m+1) \\ &= e^{-\lambda t} \frac{(\lambda t)^m}{m!}. \end{aligned}$$

This completes the proof that that the random variable N_t has Poisson distribution with parameter (λt). $\square$

Question 14. Suppose that a building has a continuum of floors indexed by the real numbers in the interval $[0, 1]$. We think of 0 as the ground floor and 1 as the building's penthouse. The building has N elevators which at any given time are independently uniformly distributed along the building's floors. You are at floor a between 0 and 1, and push a button calling for an elevator. At that time all elevators head toward you at the same speed. What is the probability that the elevator that reaches you first is traveling downward?

Answer: Assume that $a \leq \frac{1}{2}$. Denote by A the event that the elevator closest to you is below you. We need to find the probability that the elevator closest to you is above you, that is, we need to find $\mathbb{P}(A^c) = 1 - \mathbb{P}(A)$. We will compute $\mathbb{P}(A)$ by conditioning on the number of elevators

3.2. CONTINUOUS PROBABILITY

below you. Let E_n denote the event that there are exactly n elevators below you, where $0 \le n \le N$. Then,

$$\mathbb{P}(A) = \sum_{n=1}^{N} \mathbb{P}(A \cap E_n), \qquad (3.122)$$

where the summation starts at $n = 1$, as event A does not occur if no elevators are below you.

Next, we compute $\mathbb{P}(A \mid E_n)$ for $0 < n < N$, hence, assuming there is at least one elevator below as well as above you. Denote by X the distance from you to the closest elevator below you and by Y the distance from you to the closest elevator above you. For $0 < x < a$, $\{X > x\}$ is the event that all n elevators below a are in the interval $[0, a-x]$, so

$$\mathbb{P}(X > x) = \left(\frac{a-x}{a}\right)^n = \frac{1}{a^n}(a-x)^n.$$

Then, the probability density function $f_X(x)$ of X, for $x \in (0, a)$, is given by

$$\begin{aligned}
f_X(x) &= \frac{d}{dx}\mathbb{P}(X < x) = \frac{d}{dx}(1 - \mathbb{P}(X > x)) \\
&= -\frac{d}{dx}\mathbb{P}(X > x) = -\frac{d}{dx}\left(\frac{1}{a^n}(a-x)^n\right) \\
&= \frac{n}{a^n}(a-x)^{n-1}. \qquad (3.123)
\end{aligned}$$

Similarly, we have that

$$\begin{aligned}
\mathbb{P}(Y > y) &= \left(\frac{1-a-y}{1-a}\right)^{N-n} \\
&= \frac{1}{(1-a)^{N-n}}(1-a-y)^{N-n}, \quad (3.124)
\end{aligned}$$

for $0 < y < 1 - a$. Therefore, for $0 < n < N$, we obtain

$$\begin{aligned}
\mathbb{P}(A \mid E_n) &= \mathbb{P}(Y > X) \\
&= \int_0^a \mathbb{P}(Y > X \mid X = x) f_X(x)\, dx \\
&= \int_0^a \mathbb{P}(Y > x) f_X(x)\, dx.
\end{aligned}$$

From (3.123) and (3.124), we obtain that

$$\mathbb{P}(A \mid E_n) \qquad\qquad (3.125)$$
$$= \frac{n}{a^n (1-a)^{N-n}} \int_0^a (1 - a - x)^{N-n} (a - x)^{n-1}\, dx,$$

where we used the assumption that $a \leq \frac{1}{2}$.

Note that

$$\mathbb{P}(E_n) = \binom{N}{n} a^n (1-a)^{N-n}, \qquad (3.126)$$

From (3.125) and (3.126), we find that

$$\mathbb{P}(A \cap E_n) \qquad\qquad (3.127)$$
$$= \mathbb{P}(A \mid E_n) \cdot \mathbb{P}(E_n)$$
$$= n \binom{N}{n} \int_0^a (1 - a - x)^{N-n} (a - x)^{n-1}\, dx.$$

This computation assumed that $0 < n < N$, but (3.127) also holds for $n = N$. Indeed, when $n = N$, all elevators are below you, so event A must occur, that is,

$$\mathbb{P}(A \cap E_N) = \mathbb{P}(E_N) = a^N.$$

On the other hand, by letting $n = N$ in (3.127), we obtain that

$$\mathbb{P}(A \cap E_N) = N \int_0^a (a - x)^{N-1}\, dx = a^N,$$

3.2. CONTINUOUS PROBABILITY

confirming our claim.

By plugging (3.127) into (3.122), we find that

$$\mathbb{P}(A) \qquad (3.128)$$
$$= \sum_{n=1}^{N} n \binom{N}{n} \int_0^a (1-a-x)^{N-n} (a-x)^{n-1} \, dx$$
$$= \int_0^a N \sum_{n=1}^{N} \binom{N-1}{n-1} (1-a-x)^{N-n} (a-x)^{n-1} \, dx,$$

where we used the fact that

$$n \binom{N}{n} = n \cdot \frac{N!}{n!(N-n)!} = \frac{N!}{(n-1)!(N-n)!}$$
$$= N \cdot \frac{(N-1)!}{(n-1)!(N-n)!} = N \binom{N-1}{n-1}.$$

We will now use the substitution $m = n-1$ in the formula (3.128) for $\mathbb{P}(A)$. Then,

$$\mathbb{P}(A) = \int_0^a \left[N \sum_{m=0}^{N-1} \binom{N-1}{m} \times \right.$$
$$\left. \times (1-a-x)^{N-1-m} (a-x)^m \right] dx$$
$$= \int_0^a N(1-2x)^{N-1} \, dx \qquad (3.129)$$
$$= \frac{1 - (1-2a)^N}{2}, \qquad (3.130)$$

where for (3.129) we used the binomial formula

$$\sum_{m=0}^{s} \binom{s}{m} u^m v^{s-m} = (u+v)^s,$$

with $u = a-x$, $v = 1-a-x$, $s = N-1$, i.e.,

$$\sum_{m=0}^{N-1} \binom{N-1}{m} (1-a-x)^{N-1-m} (a-x)^m = (1-2x)^{N-1}.$$

We conclude that, for $a \leq \frac{1}{2}$, the probability that the elevator closest to you is above you is

$$\mathbb{P}(A^c) = \frac{1 + (1-2a)^N}{2},$$

which is precisely the probability that the elevator that reaches you first is traveling downward. Thus,

$$p_a = \frac{1 + (1-2a)^N}{2}, \qquad (3.131)$$

if $a \leq \frac{1}{2}$.

We now consider the case when $a > \frac{1}{2}$. Denote by p_a the probability that the closest elevator is above you if you are at floor a. By symmetry, it follows that

$$p_a = 1 - p_{1-a}. \qquad (3.132)$$

Since $1 - a \leq \frac{1}{2}$, we can use (3.131) and obtain that

$$p_{1-a} = \frac{1 + (2a-1)^N}{2}. \qquad (3.133)$$

From (3.132) and (3.133), we conclude that

$$p_a = \frac{1 - (2a-1)^N}{2},$$

if $\frac{1}{2} < a \leq 1$. □

Question 15. Take a wire stretched between two posts and have a large number of birds land on it at random. Take a bucket of yellow paint and, for each bird, paint the interval from it to its closest neighbor. The question is: what proportion of the wire will be painted? More precisely: as the number of birds goes to infinity, what is the limit of the expected value of the proportion of painted

3.2. CONTINUOUS PROBABILITY

wire, assuming a uniform probability distribution of birds on the wire?

Answer: Since $n \to \infty$, we can modify the problem to the one in which the wire has a shape of a circle instead of the line segment. The difference between two problems is only in the painting performed by two birds on the edge: the one that is furthest to the left and the one that is furthest to the right.

Denote by C the circle and define the function $f_n : C \to \{0, 1\}$ as follows:

$$f_n(x) = \begin{cases} 1, & \text{if the point } x \text{ is covered by paint;} \\ 0, & \text{otherwise.} \end{cases}$$

We need to find the number α defined as

$$\alpha = \lim_{n \to \infty} \mathbb{E}\left[\int_C f_n(x)\, dx\right]. \qquad (3.134)$$

The functions f_n are integrable and bounded from above by 1. Therefore, we can change the order of the integral and expected value in (3.134) and obtain that

$$\alpha = \lim_{n \to \infty} \int_C \mathbb{E}[f_n(x)]\, dx.$$

Due to symmetry, the expected value $\mathbb{E}[f_n(x)]$ does not depend on x. Hence,

$$\alpha = \lim_{n \to \infty} \mathbb{E}[f_n(X)], \qquad (3.135)$$

where X is a fixed point on the circle.

Since f_n is a function with values in the set $\{0, 1\}$, the expectation $\mathbb{E}[f_n(X)]$ is the probability of the event A_n that the point X is covered in paint. Then, from (3.135), we find that

$$\alpha = \lim_{n \to \infty} \mathbb{P}(A_n)$$

and therefore
$$\alpha = 1 - \lim_{n\to\infty} \mathbb{P}\left(A_n^C\right). \tag{3.136}$$

Thus, we set out to calculate the limit of $\mathbb{P}\left(A_n^C\right)$ as $n \to \infty$.

Denote by $\mathcal{B}_P$ the bird closest to the point X in the counter-clockwise direction and by $\mathcal{B}_Q$ the bird closest to X in the clockwise direction. Let L be the random variable that is equal to the distance (along the circle) between $\mathcal{B}_P$ and $\mathcal{B}_Q$. We will evaluate $\mathbb{P}\left(A_n^C\right)$ using the formula

$$\mathbb{P}\left(A_n^C\right) = \int_0^1 \mathbb{P}\left(A_n^C \bigg| L = t\right) F_L'(t)\, dt, \tag{3.137}$$

where F_L is the cumulative distribution function of the random variable L.

Our first task is to evaluate F_L. Denote by P the distance from the bird $\mathcal{B}_P$ to the point X. The cumulative distribution function F_L satisfies

$$\begin{aligned}F_L(t) &= 1 - \mathbb{P}(L > t) \\ &= 1 - \int_0^1 \mathbb{P}(L > t|\, P = s)\, F_P'(s)\, ds,\end{aligned} \tag{3.138}$$

where F_P is the cumulative distribution function of the random variable P. This cumulative distribution function satisfies

$$F_P(s) = 1 - \mathbb{P}(P > s) = 1 - (1-s)^n, \tag{3.139}$$

since, if the event $\{P > s\}$ occurs, then all the birds must be outside of the interval $[X, X + s]$. The derivative of $F_P(s)$ is

$$F_P'(s) = n(1-s)^{n-1}. \tag{3.140}$$

We now evaluate $\mathbb{P}(L > t|\, P = s)$. This conditional probability is equal to 1 if $s \geq t$. If $s < t$, when we condition

3.2. CONTINUOUS PROBABILITY

on the event $\{P = s\}$, the event $\{L > t\}$ has the following interpretation: Each of the $(n-1)$ independent random variables with uniform distribution on the interval $(s, 1)$ falls outside of the interval $(t-s, 1)$. Therefore, the conditional probability $\mathbb{P}(L > t | P = s)$ satisfies

$$\mathbb{P}(L > t | P = s) = \begin{cases} \left(\frac{1-t}{1-s}\right)^{n-1}, & \text{if } s < t; \\ 1, & \text{if } s \geq t. \end{cases} \quad (3.141)$$

From (3.138), (3.140), and (3.141), we obtain that

$$\begin{aligned} F_L(t) &= 1 - \int_0^t \left(\frac{1-t}{1-s}\right)^{n-1} \cdot n(1-s)^{n-1} \, ds \\ &\quad - \int_t^1 n(1-s)^{n-1} \, ds \\ &= 1 - nt(1-t)^{n-1} - (1-t)^n \end{aligned}$$

and therefore

$$F_L'(t) = n(n-1)t(1-t)^{n-2}. \quad (3.142)$$

From (3.137) and (3.142), it follows that

$$\mathbb{P}\left(A_n^C\right) = n(n-1)\int_0^1 \mathbb{P}\left(A_n^C \Big| L = t\right) t(1-t)^{n-2} \, dt. \quad (3.143)$$

We are interested in the behavior of $\mathbb{P}\left(A_n^C\right)$ as $n \to \infty$. We will split the interval of integration $[0, 1]$ into the intervals $[0, \delta]$ and $[\delta, 1]$, for some fixed positive small number δ. Let us introduce the notation

$$I_n(a, b) = \int_a^b P\left(A_n^C \Big| L = t\right) t(1-t)^{n-2} \, dt. \quad (3.144)$$

Then, the equation (3.143) can be written as

$$\mathbb{P}\left(A_n^C\right) = n(n-1)I_n(0, \delta) + n(n-1)I_n(\delta, 1). \quad (3.145)$$

For every fixed $\delta > 0$ we will prove that

$$\lim_{n \to \infty} n(n-1)I_n(\delta, 1) = 0. \qquad (3.146)$$

by using the squeeze theorem. Clearly, $n(n-1)I_n(\delta, 1)$ is positive, hence it is bounded by 0 from below. The upper bound is obtained by simply bounding the conditional probability in (3.144) by 1:

$$\begin{aligned} n(n-1)I_n(\delta, 1) &\leq n(n-1)\int_\delta^1 t(1-t)^{n-2}\, dt \\ &\leq n(n-1)\int_\delta^1 t \cdot (1-\delta)^{n-2}\, dt \\ &\leq n(n-1)\int_0^1 1 \cdot (1-\delta)^{n-2}\, dt \\ &= n(n-1) \cdot (1-\delta)^{n-2}; \end{aligned}$$

here, we used the facts that $t \leq 1$ and $1-t \leq 1-\delta$ if $\delta \leq t \leq 1$.

Since $1-\delta$ is a positive number strictly smaller than 1, the sequence $(1-\delta)^{n-2}$ converges to 0 exponentially fast in n. The quadratic function $n(n-1)$ cannot keep up with the exponential convergence to 0. The squeeze theorem now implies (3.146).

Let $\delta = \frac{1}{10}$ in (3.145). Then,

$$\mathbb{P}\left(A_n^C\right) = n(n-1)I_n\left(0, \frac{1}{10}\right) + n(n-1)I_n\left(\frac{1}{10}, 1\right).$$

Since

$$\lim_{n \to \infty} n(n-1)I_n\left(\frac{1}{10}, 1\right) = 0,$$

see (3.146), it follows that

$$P\left(A_n^C\right) = \lim_{n \to \infty} n(n-1)I_n\left(0, \frac{1}{10}\right). \qquad (3.147)$$

3.2. CONTINUOUS PROBABILITY

Recall from (3.144) that $I_n\left(0, \frac{1}{10}\right)$ is given by

$$I_n\left(0, \frac{1}{10}\right)$$
$$= \int_0^{\frac{1}{10}} \mathbb{P}\left(\left.A_n^C\right| L = t\right) t(1-t)^n \, dt. \quad (3.148)$$

The event $\{L = t\}$ means that the distance between the birds $\mathcal{B}_P$ and $\mathcal{B}_Q$ is equal to t. The bird $\mathcal{B}_P$ is the neighbor of X in the counter-clockwise direction, while the bird $\mathcal{B}_Q$ is the neighbor of X in the clockwise direction. Denote by $\mathcal{B}_{P_1}$ the neighbor of $\mathcal{B}_P$ in the counter-clockwise direction. Denote by $\mathcal{B}_{Q_1}$ the neighbor of $\mathcal{B}_Q$ in the clockwise direction. Denote by P, P_1, Q, and Q_1 the counter-clockwise distances from the birds $\mathcal{B}_P$, $\mathcal{B}_{P_1}$, $\mathcal{B}_Q$, and $\mathcal{B}_{Q_1}$ from the point X. The random variables P, P_1, Q, and Q_1 satisfy

$$0 < P < P_1 < Q_1 < Q.$$

Since we are conditioning on $\{L = t\}$, we must have

$$Q - P = 1 - t.$$

The probability of the event A_n^C conditioned on $\{L = t\}$ satisfies

$$\mathbb{P}\left(\left.A_n^C\right| L = t\right) = \mathbb{P}\left(P_1 - P \leq t, Q - Q_1 \leq t\right). \quad (3.149)$$

Note that P and Q from (3.149) are two constants such that $Q - P = 1 - t$. The random variables P_1 and Q_1 from (3.149) represent the maximum and minimum of $n-2$ independent random variables with uniform distribution on the interval (P, Q). Without loss of generality, we may assume that $P = 0$ and $Q = 1 - t$. Then, (3.149) becomes

$$\mathbb{P}\left(\left.A_n^C\right| L = t\right) = \{P_1 \leq t, Q_1 \geq 1 - 2t\}, \quad (3.150)$$

where P_1 and Q_1 from (3.150) are random variables defined as the maximum and minimum of $n-2$ independent random variables with uniform $(0, 1-t)$ distribution.

Since $t \leq \frac{1}{10}$, (3.150) can be written as

$$\mathbb{P}\left(A_n^C \middle| L = t\right) = 1 - \mathbb{P}(P_1 > t, Q_1 < 1 - 2t)$$
$$-\mathbb{P}(P_1 \leq t, Q_1 < 1 - 2t) \quad (3.151)$$
$$-\mathbb{P}(P_1 > t, Q_1 \geq 1 - 2t). \quad (3.152)$$

Due to symmetry, the probabilities in (3.151) and (3.152) are equal and therefore

$$\mathbb{P}\left(A_n^C \middle| L = t\right) = 1 - \mathbb{P}(P_1 > t, Q_1 < 1 - 2t)$$
$$-2\mathbb{P}(P_1 \leq t, Q_1 < 1 - 2t). \quad (3.153)$$

The intersection of the events $\{P_1 > t\}$ and $\{Q_1 < 1-2t\}$ is the event that all of the $n-2$ independent uniform random variables on $(1, 1-t)$ belong to the subinterval $(t, 1-2t)$. Thus,

$$\mathbb{P}(P_1 > t, Q_1 < 1 - 2t) = \left(\frac{1 - 3t}{1 - t}\right)^{n-2}. \quad (3.154)$$

Furthermore,

$$\mathbb{P}(P_1 \leq t, Q_1 < 1 - 2t)$$
$$= \int_0^t \mathbb{P}(Q_1 < 1 - 2t | P_1 = s) \cdot F'_{P_1}(s)\, ds, \quad (3.155)$$

where F_{P_1} is the cumulative distribution function of the random variable P_1. A similar argument to the one used to derive (3.139) gives us

$$F_{P_1}(s) = 1 - \mathbb{P}(P_1 > s) = 1 - \left(\frac{1 - t - s}{1 - t}\right)^{n-2},$$

and therefore

$$F'_{P_1}(s) = \frac{(n-2)(1 - t - s)^{n-3}}{(1-t)^{n-2}}. \quad (3.156)$$

3.2. CONTINUOUS PROBABILITY

We now calculate the conditional probability from (3.155). When we condition on the event $\{P_1 = s\}$, there are $n-3$ independent random variables with uniform distribution on $(s, 1-t)$. The event $\{Q_1 < 1 - 2t\}$ means that all of these $n-3$ random variables belong to the subinterval $(s, 1-2t)$. The length of the interval $(s, 1-2t)$ is $1-2t-s$, which implies that

$$\mathbb{P}\left(Q_1 < 1 - 2t \mid P_1 = s\right) = \left(\frac{1-2t-s}{1-t-s}\right)^{n-3}. \quad (3.157)$$

From (3.155), (3.156), and (3.157), we obtain that

$$\begin{aligned}
&\mathbb{P}\left(P_1 \leq t, Q_1 < 1 - 2t\right) \\
&= \int_0^t \left(\frac{1-2t-s}{1-t-s}\right)^{n-3} \cdot \frac{(n-2)(1-t-s)^{n-3}}{(1-t)^{n-2}} \, ds \\
&= \frac{n-2}{(1-t)^{n-2}} \int_0^t (1-2t-s)^{n-3} \, ds \\
&= \frac{(1-2t)^{n-2} - (1-3t)^{n-2}}{(1-t)^{n-2}}. \quad (3.158)
\end{aligned}$$

Then, from (3.153), (3.154), and (3.158) we conclude that

$$\begin{aligned}
&\mathbb{P}\left(A_n^C \mid L = t\right) \\
&= \frac{(1-t)^{n-2} - 2(1-2t)^{n-2} + (1-3t)^{n-2}}{(1-t)^{n-2}}. \quad (3.159)
\end{aligned}$$

Furthermore, we now use (3.148), (3.150), to obtain the following formula for $I_n\left(0, \frac{1}{10}\right)$.

$$\begin{aligned}
&I_n\left(0, \frac{1}{10}\right) \\
&= \int_0^{\frac{1}{10}} \frac{(1-t)^{n-2} - 2(1-2t)^{n-2} + (1-3t)^{n-2}}{(1-t)^{n-2}} \cdot \\
&\qquad \cdot t(1-t)^{n-2} \, dt \\
&= J_n(1) - 2J_n(2) + J_n(3), \quad (3.160)
\end{aligned}$$

where

$$J_n(k) = \int_0^{\frac{1}{10}} (1 - kt)^{n-2} t \, dt. \qquad (3.161)$$

From (3.136), (3.147), and (3.160), we conclude that

$$\begin{aligned}\alpha &= 1 - \lim_{n\to\infty} n(n-1)J_n(1) \\ &\quad + 2\lim_{n\to\infty} n(n-1)J_n(2) \\ &\quad - \lim_{n\to\infty} n(n-1)J_n(3). \end{aligned} \qquad (3.162)$$

We evaluate the integral in (3.161) by using the substitution $u = 1 - kt$. Then, $t = \frac{1-u}{k}$, $dt = -\frac{du}{k}$, and the bounds for integration change to $u = 1$ and $u = 1 - \frac{k}{10}$. Thus,

$$\begin{aligned}J_n(k) &= \frac{1}{k^2} \int_{1-\frac{k}{10}}^1 u^{n-2} \cdot (1-u) \, du \\ &= \frac{1}{k^2}\left(\frac{1}{n-1} - \frac{1}{n}\right) \\ &\quad - \frac{1}{k^2}\left(\frac{\left(1-\frac{k}{10}\right)^{n-1}}{n-1} - \frac{\left(1-\frac{k}{10}\right)^n}{n}\right). (3.163)\end{aligned}$$

For $k \in \{1, 2, 3\}$, the number $\left(1 - \frac{k}{10}\right)$ belongs to $(0, 1)$ and the sequences $\left(1 - \frac{k}{10}\right)^{n-1}$ and $\left(1 - \frac{k}{10}\right)^n$ converge to 0 exponentially fast. Therefore, we obtain from (3.163) that

$$\begin{aligned}\lim_{n\to\infty} n(n-1)J_n(k) &= \lim_{n\to\infty} \frac{n(n-1)}{k^2} \cdot \left(\frac{1}{n-1} - \frac{1}{n}\right) \\ &= \frac{1}{k^2}. \qquad (3.164)\end{aligned}$$

The equations (3.162) and (3.164) give us that

$$\alpha = 1 - \frac{1}{1^2} + 2 \cdot \frac{1}{2^2} - \frac{1}{3^2} = \frac{7}{18}.$$

3.2. CONTINUOUS PROBABILITY

Therefore, the limit of the expected value of the proportion of painted wire is $\frac{7}{18}$. □

Question 16. In order to generate an $N(0,1)$-distributed random number, Bob proposes the following procedure: Using the $[0,1]$-uniform random number generator, independently generate X_1, X_2, ..., X_{12}, then compute $W = X_1 + X_2 + \ldots X_{12} - 6$. Bob claims W is approximately $N(0,1)$-distributed. Could you justify his claim? Compute $E[W^4]$. What do you think of his procedure?

Answer: Bob's claim is based on a crude approximation of the standard normal distribution using the Central Limit Theorem with the 12 random variables $X_1, X_2, \ldots, X_{12}$:

$$\frac{\frac{1}{12}\sum_{i=1}^{12} X_i - \mu}{\sqrt{\frac{\sigma}{12}}} \approx N(0,1), \qquad (3.165)$$

where

$$\mu = E[X_i] = \frac{1}{2} \text{ and } \sigma = \text{var}(X_i) = \frac{1}{12} \qquad (3.166)$$

are the mean and standard deviation of the uniform distribution on $[0,1]$. Using (3.166), the approximation (3.165) becomes

$$\frac{\frac{1}{12}\sum_{i=1}^{12} X_i - \frac{1}{2}}{\sqrt{\frac{1}{12^2}}} = \sum_{i=1}^{12} X_i - 6 \approx N(0,1).$$

This is Bob's reasoning when saying that $W = \sum_{i=1}^{12} X_i - 6$ in an approximation of the standard normal distribution $N(0,1)$.

For a more accurate comparison between W and the standard normal distribution $Z \approx N(0,1)$, we will show that W and Z have the same mean, standard deviation, and skewness, while having a different kurtosis.

Let $Y_i = X_i - \frac{1}{2}$, for $i = 1:6$. Then,

$$W = \sum_{i=1}^{12} X_i - 6 = \sum_{i=1}^{12} Y_i. \qquad (3.167)$$

Denote by Y the uniform random variable on $\left[-\frac{1}{2}, \frac{1}{2}\right]$ which has the same distribution as Y_i, $i = 1:6$. The first four moments of Y are

$$E[Y] = 0; \ E[Y^2] = \frac{1}{12}; \qquad (3.168)$$

$$E[Y^3] = 0; \ E[Y^4] = \frac{1}{80}. \qquad (3.169)$$

This can be seen either by direct integration or by computing the moment generating function of Y, which is

$$E\left[e^{tY}\right] = \frac{1}{t}\left(e^{\frac{t}{2}} - e^{-\frac{t}{2}}\right),$$

and therefore

$$E\left[Y^{2p+1}\right] = 0, \ \forall \, p > 0;$$
$$E\left[Y^{2p}\right] = \frac{1}{2^{2p}(2p+1)}, \ \forall \, p \geq 0.$$

Then, from (3.167) and (3.168), we obtain that

$$E[W] = 0.$$

Moreover,

$$W^2 = \sum_{1 \leq j,k \leq 12} Y_j Y_k.$$

Thus,

$$\begin{aligned} E[W^2] &= \sum_{1 \leq j,k \leq 12} E[Y_j Y_k] \\ &= \sum_{i=1}^{12} E\left[Y_i^2\right] + \sum_{1 \leq j \neq k \leq 12} E[Y_j Y_k] \\ &= 12 E\left[Y^2\right] \\ &= 1, \end{aligned}$$

3.2. CONTINUOUS PROBABILITY

where we used the facts that $E[Y^2] = \frac{1}{12}$, see (3.168), and that, if $j \neq k$, then Y_j and Y_k are independent and therefore

$$E[Y_j Y_k] = E[Y_j] E[Y_k] = E[Y]^2 = 0.$$

To calculate the third moment of W, note that

$$W^3 = \sum_{1 \leq j,k,l \leq 12} Y_j Y_k Y_l. \qquad (3.170)$$

We will show that the expected value of every term from the sum in (3.170) is equal to 0, and therefore $E[W^3] = 0$:

If j, k, and l are all different, since Y_j, Y_k, and Y_l are independent and have mean 0, we obtain that

$$E[Y_j Y_k Y_l] = E[Y_j] E[Y_k] E[Y_l] = E[Y]^3 = 0.$$

If exactly two of j, k, and l are equal, for example, if $j = k \neq l$, then

$$E[Y_j Y_k Y_l] = E[Y_j^2 Y_l] = E[Y_j^2] E[Y_l] = E[Y^2] E[Y]$$
$$= 0,$$

since the random variables Y_j^2 and Y_l are independent and $E[Y] = 0$, see (3.168).

If $j = k = l$, then

$$E[Y_j Y_k Y_l] = E[Y_j^3] = E[Y^3] = 0.$$

Thus, we conclude that

$$E[W^3] = 0,$$

which means that W and Z have the same first three moments.

However, that will no longer be true for the fourth moment. Note that

$$W^4 = \sum_{1 \leq j,k,l,p \leq 12} Y_j Y_k Y_l Y_p. \quad (3.171)$$

Due to the independence of the random variables Y_i, $i = 1:12$, and to the facts that $E[Y] = 0$ and $E[Y^3] = 0$, the only terms from (3.171) that have nonzero expectation are the terms where all four indices are equal and the terms where two pairs of indices are equal.

Note that there are 12 terms with all four indices equal corresponding to $j = k = l = p \in \{1, 2, \ldots, 12\}$ and, according to (3.169), the expected value of each of these terms is

$$E[Y^4] = \frac{1}{80}. \quad (3.172)$$

How many terms have two pairs of indices are equal? We are looking for instances where, for example, $j = k \neq l = p$. There are $\binom{12}{2}$ ways of choosing different values for $j = k$ and $l = p$. The corresponding term $Y_j^2 Y_l^2$ appears $\binom{4}{2}$ times in the expansion of W^4 since there are $\binom{4}{2}$ ways of choosing in which of the two sums represented by W the factors Y_j would appear. Altogether, there are

$$\binom{12}{2} \cdot \binom{4}{2} = \frac{12 \cdot 11}{2} \cdot \frac{4 \cdot 3}{2} = 396$$

terms of the form $Y_j^2 Y_l^2$ in the expansion of W^4. The expected value of each of these terms is

$$E[Y_j^2 Y_l^2] = E[Y_j^2] E[Y_l^2] = \left(E[Y^2]\right)^2 = \frac{1}{144}, \quad (3.173)$$

since Y_j and Y_l are independent and $E[Y^2] = \frac{1}{12}$; see (3.168).

3.2. CONTINUOUS PROBABILITY

From (3.172) and (3.173), we conclude that

$$\begin{aligned} E[W^4] &= 12\,E[Y^4] + 396\left(E[Y^2]\right)^2 \\ &= 12\cdot\frac{1}{80} + 396\cdot\frac{1}{144} \\ &= \frac{29}{10}. \end{aligned}$$

This value is smaller than the fourth moment of the standard normal variable which is $E[Z^4] = 3$.

This procedure of trying to generate normal random variable samples is both imprecise and expensive. While the procedure has the same mean, standard deviation, and skewness as the normal random distribution, a lower kurtosis means that the tails are smaller than those of the standard normal. Further discrepancies will appear for higher moments, making this process imprecise.

Moreover, this process requires 12 uniform samples to generate one presumes sample of the standard normal distribution. That compares unfavorably with, for example, the Box–Muller method where two samples of the standard normal distribution are generated from two uniform samples. □

Question 17. There are two independently operated bus lines, both of which stop in front of your apartment building. One stops every hour, on the hour. The other also stops once an hour, but not on the hour. That is, the first bus line arrives at ..., 6am, 7am, 8am, ...; while the second bus line arrives at ..., $6 + x$ am, $7 + x$ am, $8 + x$ am, ..., where x is a positive constant. Unfortunately, you don't know the value of x. Assume that x is uniform on $(0, 1)$. What is your average waiting time?

Answer: Let W be the random variable denoting our waiting time for a bus to arrive. Without loss of generality, we can assume that we arrive at the bus stop between 6am and 7am. Denote by X the random variable with uniform distribution on $[0, 1]$ such that the arrival times for the second bus line are ..., $6 + X$, $7 + X$, ... Note that the probability density function of X is $f_X(x) = 1$, $0 \leq x \leq 1$. By conditioning on the value of X, we obtain

$$\begin{aligned} \mathbb{E}[W] &= \int \mathbb{E}[W \mid X = x] \cdot f_X(x) \, dx \\ &= \int_0^1 \mathbb{E}[W \mid X = x] \, dx. \end{aligned} \quad (3.174)$$

Given that $X = x$, for some $0 \leq x \leq 1$, we arrive at the bus stop between 6am and $6 + x$ am with probability x and wait $\frac{x}{2}$ hours on average, or we arrive at the bus stop between $6+x$ am and 7am with probability $1-x$ and wait $\frac{1-x}{2}$ hours on average. Therefore,

$$\begin{aligned} \mathbb{E}[W \mid X = x] &= x \cdot \frac{x}{2} + (1 - x) \cdot \frac{1 - x}{2} \\ &= x^2 - x + \frac{1}{2}. \end{aligned} \quad (3.175)$$

From (3.174) and (3.175), it follows that

$$\begin{aligned} \mathbb{E}[W] &= \int_0^1 \left(x^2 - x + \frac{1}{2} \right) dx \\ &= \left(\frac{x^3}{3} - \frac{x^2}{2} + \frac{x}{2} \right) \Big|_0^1 \\ &= \frac{1}{3}. \end{aligned}$$

We conclude that the average waiting time is $\frac{1}{3}$ of an hour, that is, 20 minutes. $\square$

Question 18. What is the probability that a point chosen at random from the interior of an equilateral triangle is closer to the center than to any of its sides?

3.2. CONTINUOUS PROBABILITY

Answer: Let $\triangle ABC$ be the equilateral triangle, draw the perpendiculars from the vertices A, B, and C onto the opposite sides, and denote their intersection point by O. The equilateral triangle is split into six congruent triangles. Due to symmetry, the probability that a point chosen at random from the interior of the equilateral triangle $\triangle ABC$ is closer to its center O than to any of its sides is equal to the probability that a point chosen at random from the interior of one of these six congruent triangles, such as $\triangle OA_1C$, see the figure below, is closer to the center O than any of its sides, which in this case is side A_1C.

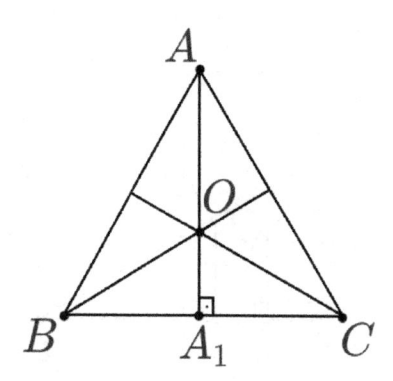

Denote by l the length of the side of the equilateral triangle $\triangle ABC$. Then,

$$A_1C = \frac{l}{2}; \quad OC = \frac{l\sqrt{3}}{3}; \quad OA_1 = \frac{l\sqrt{3}}{6}, \qquad (3.176)$$

and the area of the triangle $\triangle OA_1C$ is

$$\text{Area}(\triangle OA_1C) = \frac{OA_1 \cdot A_1C}{2} \stackrel{3.176}{=} \frac{l^2\sqrt{3}}{24}. \qquad (3.177)$$

We now address the question of which points from the triangle $\triangle OA_1C$ are closer to the side A_1C than to the point O? For simplicity, assume that we use rectangular coordinates with the vertex A_1 being the center $(0,0)$, the x–axis along the side A_1C, and the y–axis along the side A_1O. Then, the vertex O has coordinates $\left(0, \frac{l\sqrt{3}}{6}\right)$ and the vertex C has coordinates $\left(\frac{l}{2}, 0\right)$.

To begin with, let A_2 be the point on the side OA_1 which is equidistant to the point O and to the side A_1C. Then, A_2 is the midpoint of the segment OA_1 and has coordinates $\left(0, \frac{l\sqrt{3}}{12}\right)$; see also the figure below.

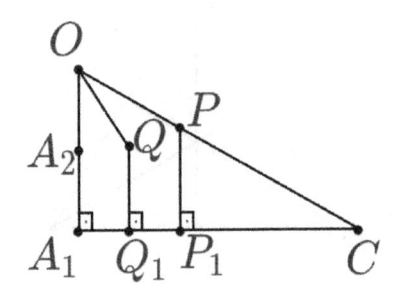

Furthermore, let P be the point on the side OC which is equidistant to the point O and to the side A_1C. Draw the perpendicular line from P onto A_1C and let P_1 be the point on A_1C such that $PP_1 \perp A_1C$. If we denote by (x_P, y_P) the coordinates of the point P, then the distance from P to A_1C is equal to $PP_1 = y_P$ and $OP = PP_1 = y_P$. Thus,

$$PC = OC - OP \stackrel{3.176}{=} \frac{l\sqrt{3}}{3} - y_P. \qquad (3.178)$$

3.2. CONTINUOUS PROBABILITY

Since $\angle PCP_1 = \angle OCA_1 = 30°$, it follows that its sine satisfies $\sin(\angle PCP_1) = \frac{1}{2}$ and therefore

$$\sin(\angle PCP_1) = \frac{PP_1}{PC} = \frac{y_P}{\frac{l\sqrt{3}}{3} - y_P} = \frac{1}{2}. \quad (3.179)$$

By solving (3.179) for y_P, we obtain that

$$y_P = \frac{l\sqrt{3}}{9}. \quad (3.180)$$

Moreover, since $\tan(\angle PCP_1) = \tan(30°) = \frac{1}{\sqrt{3}}$, we obtain that

$$\tan(\angle PCP_1) = \frac{PP_1}{CP_1} = \frac{y_P}{CP_1} = \frac{1}{\sqrt{3}}.$$

Thus,

$$CP_1 = y_P\sqrt{3} \overset{3.180}{=} \frac{l}{3} \quad (3.181)$$

and, using (3.176) and (3.181), we find that

$$x_P = A_1 P_1 = A_1 C - CP_1 = \frac{l}{2} - \frac{l}{3} = \frac{l}{6}. \quad (3.182)$$

It is now important to note that all the points from the triangle $\triangle PP_1C$ are closer to the side A_1C than to the point O: The distance between any point inside the triangle $\triangle PP_1C$ and the side A_1C is smaller than PP_1, while the distance between a point inside the triangle $\triangle PP_1C$ and the point O is greater than OP. Since $PP_1 = OP$, we conclude that all the points from the triangle $\triangle PP_1C$ are closer to the side A_1C than to the point O.

Moreover, from (3.180) and (3.181), we obtain that the area of the triangle $\triangle PP_1C$ is

$$\begin{aligned} \text{Area}(\triangle PP_1C) &= \frac{PP_1 \cdot CP_1}{2} = \frac{y_P \cdot CP_1}{2} \\ &= \frac{l^2\sqrt{3}}{54}. \end{aligned} \quad (3.183)$$

Thus, any points from the triangle $\triangle OA_1C$ that are closer to the point O than to the side A_1C are located in the right trapezoid A_1P_1PO. Let Q be a point with the property that the distance from Q to the side A_1C is exactly equal to the distance OQ. Denote by (x_Q, y_Q) the coordinates of Q. Note that

$$0 \le x_Q \le x_P \stackrel{3.182}{=} \frac{l}{6},$$

since Q is inside the right trapezoid A_1P_1PO.

The distance OQ between the point $Q(x_Q, y_Q)$ and the point $O(0, \frac{l\sqrt{3}}{6})$ is

$$OQ = \sqrt{x_Q^2 + \left(\frac{l\sqrt{3}}{6} - y_Q\right)^2}.$$

The distance from Q to the side A_1C is y_Q. Thus, the distance from Q to A_1C is equal to OQ if and only if

$$y_Q = \sqrt{x_Q^2 + \left(\frac{l\sqrt{3}}{6} - y_Q\right)^2}$$

$$\iff y_Q^2 = x_Q^2 + \left(\frac{l\sqrt{3}}{6} - y_Q\right)^2$$

$$\iff y_Q^2 = x_Q^2 + \frac{l^2}{12} - \frac{l\sqrt{3}}{3}y_Q + y_Q^2$$

$$\iff \frac{l\sqrt{3}}{3}y_Q = x_Q^2 + \frac{l^2}{12}$$

$$\iff y_Q = \frac{x_Q^2 \sqrt{3}}{l} + \frac{l\sqrt{3}}{12}.$$

Since $A_1P_1 = \frac{l}{6}$, see (3.182), we conclude that all the points inside the right trapezoid A_1P_1PO which are closer to the side A_1C than to the point O belong to the set

$$S_{A_1C} = \left\{(x,y) \ : \ 0 \le x \le \frac{l}{6}; \ 0 \le y \le \frac{x^2\sqrt{3}}{l} + \frac{l\sqrt{3}}{12}\right\}.$$

3.2. CONTINUOUS PROBABILITY

The area of S_{A_1C} is given by

$$\begin{aligned}
\text{Area}(S_{A_1C}) &= \int_0^{\frac{l}{6}} \int_0^{\frac{x^2\sqrt{3}}{l}+\frac{l\sqrt{3}}{12}} 1\, dy\, dx \\
&= \int_0^{\frac{l}{6}} \frac{x^2\sqrt{3}}{l} + \frac{l\sqrt{3}}{12}\, dx \\
&= \left(\frac{x^3\sqrt{3}}{3l} + \frac{x \cdot l\sqrt{3}}{12}\right)\Bigg|_0^{\frac{l}{6}} \\
&= \frac{l^2\sqrt{3}}{648} + \frac{l^2\sqrt{3}}{72} \\
&= \frac{l^2 \cdot 5\sqrt{3}}{324}. \quad\quad (3.184)
\end{aligned}$$

From (3.177), (3.183), and (3.184), we conclude that the probability that a point inside the triangle $\triangle OA_1C$ is closer to the side A_1C than to the point O is equal to

$$\begin{aligned}
&\frac{\text{Area}(S_{A_1C}) + \text{Area}(\triangle PP_1C)}{\text{Area}(\triangle OA_1C)} \\
&= \left(\frac{l^2 \cdot 5\sqrt{3}}{324} + \frac{l^2\sqrt{3}}{54}\right) \div \frac{l^2\sqrt{3}}{24} \\
&= \left(\frac{5}{324} + \frac{1}{54}\right) \cdot 24 \\
&= \frac{22}{27}. \quad\quad (3.185)
\end{aligned}$$

Recall that, due to symmetry, the probability that a point chosen at random from the interior of the equilateral triangle $\triangle ABC$ is closer to one of its sides than to its center O is equal to the probability that a point chosen at random from the interior of triangle $\triangle OA_1C$ is closer to the side A_1C than to the point O.

Thus, from (3.185), we conclude that the probability that a point chosen at random from the interior of an equilateral triangle is closer to the center than to any of its sides is

$$1 - \frac{22}{27} = \frac{5}{27}. \quad \square$$

Question 19. In the segment $[0,1]$, n points are chosen uniformly at random. For every point, one of the two directions (left or right) is chosen randomly and independently. At the same moment in time all n points start moving in the chosen direction with speed 1. The collisions of all points are elastic. That means, after two points bump into each other, they start moving in the opposite directions with the same speed of 1. When a point reaches an end of the segment it sticks to it and stops moving. Find the expected time when the last point sticks to the end of the segment.

Answer: Let us put a hat on each of the points. Assume that once two points collide and change directions, they exchange the hats. The hats are always traveling in the same direction. The hats are moving for as long as the points are moving. Therefore, we need to determine the expected time when the last hat sticks to the end of the segment. For each $i \in \{1, 2, \ldots, n\}$, denote by T_i the time until the hat i stops moving. Clearly, the random variables $T_1, T_2, \ldots, T_n$ are independent and with identical distribution. Observe also that T_i is uniform on $[0,1]$. We need to find out the expected value of the random variable

$$M = \max\{T_1, T_2, \ldots, T_n\}.$$

We will use the following formula for expected value of a non–negative continuous random variable:

$$\mathbb{E}[M] = \int_0^{+\infty} \mathbb{P}(M > t) \, dt. \qquad (3.186)$$

Note that $\mathbb{P}(M \leq t)$ is very easy to calculate. First, it is equal to 1 if $t \geq 1$. For $t \in (0,1)$, the event $\{M \leq t\}$ is the

3.2. CONTINUOUS PROBABILITY

same as the intersection of the events $\{T_1 \leq t\}$, $\{T_2 \leq t\}$, ..., $\{T_n \leq t\}$. The probability of each such event is t and their independence implies that

$$\mathbb{P}(M \leq t) = \mathbb{P}(T_1 \leq t) \cdots \mathbb{P}(T_n \leq t) = t^n.$$

and therefore

$$\mathbb{P}(M > t) = 1 - t^n. \qquad (3.187)$$

From (3.186) and (3.187), we obtain that

$$\begin{aligned}
\mathbb{E}[M] &= \int_0^1 (1 - t^n)\, dt \\
&= \left(t - \frac{t^{n+1}}{n+1} \right) \bigg|_{t=0}^{t=1} \\
&= 1 - \frac{1}{n+1} \\
&= \frac{n}{n+1}.
\end{aligned}$$

We conclude that the expected time when the last of the n points stops moving is $\frac{n}{n+1}$. □

Question 20. There are n points uniformly distributed in a unit disk. What is the median of the smallest distance between these points and the center of the disk?

Answer: We may assume that the center of the unit disk is at the origin. Denote the points uniformly distributed in it by X_i, $i = 1:n$. The question is asking for the value of r such that

$$\mathbb{P}\left(\min_i \|X_i\| > r \right) = \frac{1}{2}, \qquad (3.188)$$

where $\|X_i\|$ denotes the distance between the point X_i and the origin. Since X_i's are independent, we have

$$\mathbb{P}\left(\min_i \|X_i\| > r\right) = \prod_{i=1}^{n} \mathbb{P}\left(\|X_i\| > r\right)$$
$$= \prod_{i=1}^{n} \left(1 - \mathbb{P}\left(\|X_i\| < r\right)\right). \quad (3.189)$$

For $1 \leq i \leq n$, since X_i is uniformly distributed inside the unit disk, it follows that $\|X_i\| < r$ if and only if the point X_i is in the disk with center at origin and radius r, for $0 \leq r \leq 1$. Hence,

$$\mathbb{P}\left(\|X_i\| < r\right) = \frac{\pi r^2}{\pi} = r^2. \quad (3.190)$$

From (3.189) and (3.190), we obtain that

$$\mathbb{P}\left(\min_i \|X_i\| > r\right) = \left(1 - r^2\right)^n. \quad (3.191)$$

From (3.188) and (3.191), it follows that the median of the smallest distance between the points and the center of the disk is the solution to

$$\left(1 - r^2\right)^n = \frac{1}{2}.$$

Thus,

$$r = \left(1 - \left(\frac{1}{2}\right)^{\frac{1}{n}}\right)^{\frac{1}{2}}. \quad \square$$

Chapter 4

Combinatorial Puzzles

4.1 Counting Challenges

Question 1. Is it true that any set of ten distinct numbers between 1 and 100 contains two disjoint nonempty subsets with the same sum?

Answer: Let S be a set of ten distinct numbers between 1 and 100. Then, the sum of the elements of S is between

$$\sum_{i=1}^{10} i = 55 \quad \text{and} \quad \sum_{i=91}^{100} i = 955.$$

Each of $2^{10} - 1 = 1023$ non-empty subsets of S has a sum between 1 and 955. Since $1023 > 955$, we can use the pigeonhole principle. The pigeonhole principle is a name for the following obvious proposition:

If n pigeons are placed into m holes, and if $n > m$, then there is a hole that contains at least two pigeons.

In this problem, the holes correspond to the possible sums of elements. The pigeons correspond to the subsets. The pigeonhole principle implies that there are two subsets with equal sums of elements. In other words, there are

216   CHAPTER 4. COMBINATORIAL PUZZLES

two non-empty subsets of S, say A and B, not necessarily disjoint, such that

$$\sum_{i \in A} i = \sum_{i \in B} i.$$

Then, by removing the common elements in $A \cap B$, we obtain $A \setminus B$ and $B \setminus A$, two disjoint non-empty subsets with the same sum.   □

Question 2. In how many ways, counting ties, can four horses cross the finishing line?

(For example, two horses can finish in three ways: A wins, B wins, A and B tie.)

Answer: *First Solution:* We are going to count in how many ways the four horses cross the finishing line depending on the number of ties.

• If there are no ties, the total number of race outcomes is $4! = 24$.

• If there is exactly one tie, there are $\binom{4}{2} = 6$ possible ways for two horses to tie. After that, there are three different slots for the other two horses and the two horses who tie to finish in, and there are $3! = 6$ orderings of these three finishing slots. Overall, there are $6 \cdot 6 = 36$ possible race outcomes with one tie.

• If there are two ties, there will be one winning pair of horses which is followed by the second pair of horses who are ties. Thus, the number of race outcomes in this case is equal to the number of possibilities of choosing a winning pair out of four horses which is $\binom{4}{2} = 6$.

• If there is one tie with three horses, there are $\binom{4}{3} = 4$ possible ways for three horses to tie. For each three-way tie, there are two possible race outcomes: either the three horses that tie win the race, followed by the fourth horse,

4.1. COUNTING CHALLENGES

or the three horses that tie end up in second place and the race is won by the fourth horse. Overall, there are $4 \cdot 2 = 8$ possible race outcomes in this case.

• If all four horses end up tied, this corresponds to 1 race outcome.

We conclude that the total number of possible outcomes for the race are

$$24 + 36 + 6 + 8 + 1 = 75.$$

Second Solution: Name the four horses A, B, C, and D. Each horse race result can be thought of as a partition of the set $\{A, B, C, D\}$ of horses into some number of non-empty blocks, followed by the ordering of the blocks according to their finish time. The order of the horses within the same block does not matter as those horses finish the race at the same time. For example, one horse race result that we want to count is: (First place) A and C, (Second place) B and D; which can be thought of as a partition of $\{A, B, C, D\}$ into two non-empty blocks $\{A, C\}$ and $\{B, D\}$, followed by the ordering of the blocks so that horses A and C are tied for the first place, and horses B and D are tied for the second place.

Denote by $S(n, k)$ the number of partitions of n distinct objects into k non-empty blocks, where the order of the blocks does not matter. Then, we conclude that the total number of different race results is

$$S(4, 4) \cdot 4! + S(4, 3) \cdot 3! + S(4, 2) \cdot 2! + S(4, 1) \cdot 1!,$$

since once the four horses are partitioned into k non-empty blocks, $k \in \{1, 2, 3, 4\}$, the blocks themselves can be ordered in $k!$ ways.[1]

---

[1] The numbers $S(n, k)$ are known as the sequence of Stirling numbers of the second kind in enumerative combinatorics. Although there is no simple explicit formula for $S(n, k)$, the Stirling

Note that $S(4,1) = 1$, as $\{A, B, C, D\}$ is the only partition of $\{A, B, C, D\}$ into 1 block and $S(4,4) = 1$, as $\{\{A\}, \{B\}, \{C\}, \{D\}\}$ is the only partition of $\{A, B, C, D\}$ into 4 blocks. Moreover, $S(4,2) = 7$, since $\{\{A, B, C\}, \{D\}\}$, $\{\{A, B, D\}, \{C\}\}$, $\{\{A, C, D\}, \{B\}\}$, $\{\{B, C, D\}, \{A\}\}$, $\{\{A, B\}, \{C, D\}\}$, $\{\{A, C\}, \{B, D\}\}$, $\{\{A, D\}, \{B, C\}\}$ are all the partitions of $\{A, B, C, D\}$ into 2 blocks. Finally, $S(4,3) = \binom{4}{2} = 6$, since a partition of $\{A, B, C, D\}$ into 3 blocks is uniquely determined by a choice of which two horses end up in the same block (that is, finish the race in a tie). Plugging these values in, we conclude that the total number of different race results is

$$1 \cdot 4! + 6 \cdot 3! + 7 \cdot 2! + 1 \cdot 1! = 24 + 36 + 14 + 1 = 75. \quad \square$$

Question 3. For a 6×4 rectangle, the square in the second row and third column is painted black. How many rectangles with sides along the grid lines contain the black square?

Answer: Label the vertical grid lines V_1 through V_7, starting with the line containing the leftmost side of the 6×4 rectangle, and finishing with the line containing its rightmost side. Similarly, label the horizontal grid lines H_1 through H_5, starting with the line containing the bottom side of the 6×4 rectangle, and finishing with the line containing its top side. Every rectangle contained within the 6×4 rectangle and with sides along the grid lines, is uniquely determined by a pair of vertical lines containing its vertical sides, and by a pair of horizontal lines containing its horizontal sides. Such a rectangle will contain the black square if and only if the vertical line containing

---

numbers of the second kind can be calculated via the recurrence $S(n, k) = S(n - 1, k - 1) + kS(n - 1, k)$ which can be obtained by classifying the partitions into two types depending on whether object n is in a block by itself or not.

4.1. COUNTING CHALLENGES

its leftmost side belongs to $\{V_1, V_2, V_3\}$, the vertical line containing its rightmost side belongs to $\{V_4, V_5, V_6, V_7\}$, the horizontal line containing its bottom side belongs to $\{H_1, H_2\}$, and the horizontal line containing its top side belongs to $\{H_3, H_4, H_5\}$. Therefore, the number of such rectangles is $3 \cdot 4 \cdot 2 \cdot 3 = 72$. □

Question 4. 30 senators attend a session. Each senator is an enemy to exactly 10 other senators. In how many ways can one form a 3-member committee so that either all of the senators on the committee are mutual enemies, or no two of the senators on the committee are enemies?

Answer: The total number of 3-member committees is $\binom{30}{3} = 4060$. A 3-member committee such that either all of the senators on the committee are mutual enemies, or no two of the senators on the committee are enemies, will be called *consistent*. Note that every 3-member *inconsistent* committee contains exactly 2 senators, who both have one friend and one enemy on that committee. Since each senator is an enemy to exactly 10 senators, and thus a friend to exactly 19 senators, we deduce that each senator belongs to precisely $10 \cdot 19 = 190$ *inconsistent* 3-member committees. Note that each *inconsistent* 3-member committee has two members who each have one enemy and one friend on the committee. Hence, the number of *inconsistent* 3-member committees is $\frac{30 \cdot 190}{2} = 2850$. Therefore, the total number of *consistent* 3-member committees is

$$\binom{30}{3} - \frac{30 \cdot 190}{2} = 4060 - 2850 = 1210. \quad \Box$$

Question 5. In a chess tournament, each two players played each other once. Each player got 1 point for a win, 1/2 point for a draw, and 0 points for a loss. Let S be the

220   CHAPTER 4. COMBINATORIAL PUZZLES

set of the 10 lowest-scoring players. If every player got exactly half his total score from the games played against the players from the set S, how many players were in the tournament?

Answer: Suppose there were $n + 10$ players in the tournament overall. There are $\binom{n}{2}$ games played among the n players not in S, and there are $\binom{n}{2}$ points earned in these games, as the outcome of any single game distributes one point among the two players playing the game. Since every player got exactly half his total score from the games played against the players from S, this implies that the n players not in S also earned $\binom{n}{2}$ points while playing against the players in S. Similarly, the 10 players in S played $\binom{10}{2} = 45$ games among themselves and, thus, earned 45 points playing each other. Again, since every player got exactly half his total score from the games played against the players from S, the 10 players in S also earned 45 points playing against the n players not in S. Therefore, the total number of points earned in this tournament is

$$2 \cdot \binom{n}{2} + 2 \cdot 45 \;=\; n^2 - n + 90. \qquad (4.1)$$

On the other hand, there are $\binom{n+10}{2}$ games played overall in the tournament, and with one point being earned per game, the total number of points earned in the tournament can also be expressed as

$$\binom{n+10}{2} \;=\; \frac{(n+10)(n+9)}{2}. \qquad (4.2)$$

From (4.1) and (4.2), we obtain that

$$n^2 - n + 90 \;=\; \frac{(n+10)(n+9)}{2},$$

4.1. COUNTING CHALLENGES

which is equivalent to

$$n^2 - 21n + 90 = (n-6)(n-15) = 0.$$

At this point, it seems as if there are two possible values for n, namely, $n = 6$ and $n = 15$. However, note that the top-scoring n players got $2 \cdot \binom{n}{2} = n(n-1)$ points in total, for an average of $n-1$ points. The 10 lowest-scoring players got $2 \cdot \binom{10}{2} = 90$ points in total, for an average of 9 points. Thus, we must have $n - 1 \geq 9$, that is, $n \geq 10$. Therefore, n must be equal to 15, and our tournament had $n + 10 = 25$ players. □

Question 6. The desks in a classroom with 39 desks are arranged in a rectangular 3×13 grid. In the morning, 39 students take seats, one per desk. After recess, the students take seats again, one per desk, in such a way that each student sits at a desk that is horizontally or vertically adjacent to the desk where they sat from before the recess. How many different seating arrangements are there after recess?

Answer: There are zero such arrangements! Color the 39 desks, arranged in a rectangular 3×13 grid, black and white, using the checkerboard coloring. There are either 20 black desks and 19 white desks, or 19 black desks and 20 white desks. Since after recess, each of the 39 students sits at a desk that is horizontally or vertically adjacent to the desk where they sat before the recess, each student sitting at a white desk before the recess would have to sit at a black desk after the recess, and vice versa. However, this is impossible as the number of black desks is different from the number of white desks. □

Question 7. The desks in a classroom with 24 desks are arranged in a rectangular 2×12 grid. In the morning,

24 students take seats, one per desk. After recess, the students take seats again, one per desk, in such a way that each student sits at a desk that is horizontally or vertically adjacent to the desk where they sat from before the recess. How many different seating arrangements are there after recess?

Answer: Denote by α_n the number of different arrangements of ordered pairs (i,j) with $i \in \{1, 2, \ldots, n\}$ and $j \in \{1,2\}$ in a $2 \times n$ grid such that the coordinates (i',j') of the pair (i,j) satisfy

$$|i' - i| + |j' - j| = 1.$$

We need to calculate α_{12}. Let us paint the cells of the grid in a checkerboard pattern: the cells with coordinates (i,j) are white if $i + j$ is even, and black if $i + j$ is odd. The ordered pairs (k,l) for which $k + l$ is even have to be placed in black cells. The ordered pairs whose sum of coordinates is odd have to be placed in white cells. Denote by β_n the number of ways in which the black cells can be filled with ordered pairs. Clearly, the number of ways to fill the white cells is also β_n and the number α_n satisfies

$$\alpha_n = \beta_n^2.$$

Let us now find the recursive relation for β_n by studying the ways in which the ordered pairs (k,l) with $2 \mid (k+l)$ can be placed on black cells. The pair $(1,1)$ can be placed either on cell $(1,2)$ or on cell $(2,1)$. If it is placed on $(1,2)$, then the number of ways to fill the remaining black cells is β_{n-1}. If the pair $(1,1)$ is placed on the cell $(2,1)$, then the cell $(1,2)$ must contain $(2,2)$. The remaining black cells can be filled in β_{n-2} ways. Therefore,

$$\beta_n = \beta_{n-1} + \beta_{n-2}.$$

The initial terms β_1 and β_2 satisfy $\beta_1 = 1$ and $\beta_2 = 2$. The sequence $(\beta_n)_{n=1}^{\infty}$ is a shifted Fibonacci sequence (the

4.1. COUNTING CHALLENGES

conventional Fibonacci sequence has $\beta_1 = \beta_2 = 1$). The first 12 numbers are

$$\beta_1 = 1,\ \beta_2 = 2,\ \beta_3 = 3,\ \beta_4 = 5,$$
$$\beta_5 = 8,\ \beta_6 = 13,\ \beta_7 = 21,\ \beta_8 = 34$$
$$\beta_9 = 55,\ \beta_{10} = 89,\ \beta_{11} = 144,\ \beta_{12} = 233.$$

Therefore, $\alpha_{12} = 233^2 = 54289$. □

Question 8. How many 8×8 matrices with all entries 0 or 1 are there, such that the sum of each row and each column is odd?

Answer: Let us denote by E the matrix of the format 8×8. Let us consider the 7×7 sub matrix S obtained by removing the last row and the last column. There are 2^{49} ways to put the numbers in this sub matrix S. Each configuration of numbers in S uniquely corresponds to one 8×8 matrix E that satisfies the required conditions. Indeed, let us denote by $c_1, c_2, \ldots, c_7$ the sums of numbers in the seven columns of S. Similarly, denote by $r_1, r_2, \ldots, r_7$ the sums of numbers in the seven rows of S. The first 7 numbers in the last row of E should be $(1 - c_1)$, $(1 - c_2)$, ..., $(1 - c_7)$, where the calculations are done modulo 2. The first 7 numbers of the last column of E must be $(1 - r_1), (1 - r_2), \ldots, (1 - r_7)$. The number x in the bottom-right corner of E must satisfy the system of congruencies

$$(1 - c_1) + \cdots + (1 - c_7) + x \equiv 1 \pmod{2} \text{ and} \quad (4.3)$$
$$(1 - r_1) + \cdots + (1 - r_7) + x \equiv 1 \pmod{2}. \quad (4.4)$$

Note that the numbers $C = c_1 + \cdots + c_7$ and $R = r_1 + \cdots + r_7$ are the sums of all entries in the matrix S and therefore are equal. Thus, $C = R$ and the system of congruencies (4.3)–(4.4) has a solution $x \in \{0, 1\}$. Therefore, we conclude that there are 2^{49} matrices E with the required property. $\square$

Question 9. A perfect in–shuffle of a deck of 52 cards is defined as follows: The deck is cut in half followed by interleaving of the two piles. So, if the cards were labeled 0, 1, 2, ..., 51, the new sequence is 0, 26, 1, 27, 2, 28, ... With repeated in–shuffles, do we ever get back the original order, and if so, after how many in–shuffles?

Answer: The answer is yes. The perfect in-shuffle is a permutation. There are 52! permutations of the set $\{0, 1, \ldots, 51\}$. The set of those permutations forms a group $\mathbb{S}_{52}$ with respect to the composition of permutations. This group is, obviously, finite. It has 52! elements. Every element ρ in the finite group is of finite order. In other words, for each element ρ there exists an integer k such that ρ^k is the identity.

We now consider the perfect in-shuffle. Let us denote it by σ. We will prove that the order of σ is 8. In other words, we will prove that σ^8 is the identity, and σ^k is not the identity for $k < 8$.

The fixed points of σ are 0 and 51. These fixed points form cycles of length 1. The remaining cycles of the perfect in-shuffle σ are:

$$\begin{aligned} C_1 &= (1, 26, 13, 32, 16, 8, 4, 2) \\ C_2 &= (3, 27, 39, 45, 48, 24, 12, 6) \\ C_3 &= (5, 28, 14, 7, 29, 40, 20, 10) \\ C_4 &= (9, 30, 15, 33, 42, 21, 36, 18) \\ C_5 &= (11, 31, 41, 46, 23, 37, 44, 22) \\ C_6 &= (17, 34) \\ C_7 &= (19, 35, 43, 47, 49, 50, 25, 38) \, . \end{aligned}$$

4.1. COUNTING CHALLENGES

The permutation σ can be written as the composition

$$\sigma = C_1 \circ C_2 \circ C_3 \circ C_4 \circ C_5 \circ C_6 \circ C_7.$$

These cycles have three different lengths: 1, 2, or 8. Since the least common multiplier of the numbers 1, 2, and 8 is 8, the perfect in-shuffle σ is a permutation of order 8. Therefore, when σ is applied eight times, the resulting permutation is the identity. $\square$

Question 10. In how many ways can you divide 7 candies and 14 stickers among 4 children such that each child gets at least one candy and also gets more stickers than candies?

Answer: First, we distribute the candies. Let c_i denote the number of candies the ith child gets, $i \in \{1, 2, 3, 4\}$. Counting the ways to distribute 7 candies among 4 children, so that each child gets at least one candy, is then equivalent to counting positive integer solutions to equation

$$c_1 + c_2 + c_3 + c_4 = 7.$$

Any such solution, say $(c_1, c_2, c_3, c_4) = (1, 2, 3, 1)$ can be uniquely represented as a sequence

$$\star \,|\, \star \star \,|\, \star \star \star \,|\, \star$$

of 7 stars (representing candies) and 3 bars (separating the candy lots that 4 children receive). So, the number of positive integer solutions to equation $c_1 + c_2 + c_3 + c_4 = 7$ is precisely the number of sequences of 7 stars and 3 bars, where bars are placed in between the stars, but not before the first or after the last star, and so that no place between two stars can be occupied by more than one bar (since each child gets at least one candy). Hence, one only needs to choose 3 out of the 6 places between the stars and place bars there. Thus, the number of ways to distribute

7 candies among 4 children, so that each child gets at least one candy, is $\binom{6}{3}$.

Once we have distributed the candies, we need to distribute the stickers so that each child gets more stickers than candies. Let s_i denote the number of stickers the ith child gets, $i = 1, 2, 3, 4$. Counting integer solutions to equation $s_1 + s_2 + s_3 + s_4 = 14$ such that $s_i > c_i$ is equivalent to counting positive integer solutions to equation $(s_1 - c_1) + (s_2 - c_2) + (s_3 - c_3) + (s_4 - c_4) = 14 - 7 = 7$. Thus, setting $d_i = s_i - c_i$, we need to count positive integer solutions to equation $d_1 + d_2 + d_3 + d_4 = 7$, which is precisely the counting problem we have already solved in the previous paragraph. Thus, (with candies already distributed among the children) the number of ways to distribute the stickers so that each child gets more stickers than candies is $\binom{6}{3}$.

We conclude that the number of ways in which we can divide 7 candies and 14 stickers among 4 children, so that each child gets at least one candy and also gets more stickers than candies, is

$$\binom{6}{3}^2 = 400. \quad \square$$

Question 11. Let A and B be two sets with the property that there are exactly 144 sets that are subsets of at least one of A or B. How many elements are in the union of A and B?

Answer: Denote by x the number of elements in $A \setminus B$, by y the number of elements in $B \setminus A$, and by z the number of elements in $A \cap B$. We want to find $x + y + z$.

The number of subsets of A is 2^{x+z}, the number of subsets of B is 2^{y+z}, and the number of subsets of both

4.1. COUNTING CHALLENGES

A and B, that is, of $A \cap B$, is 2^z. If we consider all the 2^{x+z} subsets of A, add to them the 2^{y+z} subsets of B, and then eliminate all the duplicates (which are exactly the 2^z subsets of $A \cap B$), we obtain all the sets that are subsets of at least one of A and B. Thus, the number of such subsets is equal to $2^{x+z} + 2^{y+z} - 2^z$, and since we were told that there are exactly 144 subsets with this property, we obtain that

$$2^{x+z} + 2^{y+z} - 2^z = 144.$$

Then,

$$2^z \left(2^x + 2^y - 1\right) = 16 \cdot 9,$$

and we conclude that $2^z = 16$ and $z = 4$ and $2^x + 2^y - 1 = 9$ and therefore $2^x + 2^y = 10$. The latter equation has only two solutions, namely $(x, y) = (1, 3)$ and $(x, y) = (3, 1)$. Either way, $x + y + z = 8$ and we conclude that there are 8 elements in $A \cup B$. □

Question 12. Show that the number of ways of writing 2020 as a sum of distinct positive integers is equal to the number of ways of writing 2020 as a sum of positive odd integers.

Answer: We will prove that the claim holds for every nonnegative integer n, not only for $n = 2020$. For every nonnegative integer n, denote by $p_\text{d}(n)$ the number of ways of writing n as a sum of distinct positive integers, and denote by $p_\text{odd}(n)$ the number of ways of writing n as a sum of positive odd integers. In order to show that

$$p_\text{d}(n) = p_\text{odd}(n)$$

for every nonnegative integer n, it suffices to show that the generating functions

$$\text{D}(x) = \sum_{n \geq 0} p_\text{d}(n) \cdot x^n$$

and
$$\text{Odd}(x) = \sum_{n \geq 0} p_{\text{odd}}(n) \cdot x^n$$
of the two sequences $\{p_{\text{d}}(n)\}_{n \geq 0}$ and $\{p_{\text{odd}}(n)\}_{n \geq 0}$ are the same!

We first show that
$$\begin{aligned} \text{D}(x) &= (1+x^1) \cdot (1+x^2) \cdot (1+x^3) \cdot (1+x^4) \cdots \\ &= \prod_{k \geq 1} (1+x^k). \end{aligned} \quad (4.5)$$

Indeed, the term $p_{\text{d}}(n) \cdot x^n$ in $\text{D}(x)$ on the left-hand side of (4.5) is the sum of the products of terms coming from each of the parentheses on the right-hand side of (4.5); the term from the kth parenthesis $(1+x^k)$ being either x^k or $x^0 = 1$, depending on whether the positive integer k is used once or not at all when writing n as a sum of distinct positive integers.

Similarly,
$$\begin{aligned} \text{Odd}(x) &= (1 + x^1 + x^2 + \ldots) \times \\ &\quad \times (1 + x^3 + x^6 + \ldots) \times \\ &\quad \times (1 + x^5 + x^{10} + \ldots) \times \cdots \quad (4.6) \\ &= \prod_{k \geq 1,\ k \text{ odd}} \frac{1}{(1-x^k)}, \quad (4.7) \end{aligned}$$

the last line following from summing up the geometric series in each of the parenthesis on the right-hand side of (4.6). Indeed, the term $p_{\text{odd}}(n) \cdot x^n$ in $\text{Odd}(x)$ on the left-hand side of (4.6) is the sum of the products of terms coming from each of the parentheses on the right-hand side of (4.6); the term from the parenthesis $(1 + x^k + x^{2k} + \ldots)$ being $x^{i_k k}$ if the positive odd integer k is used i_k times when writing n as a sum of positive odd integers.

4.1. COUNTING CHALLENGES

In order to show that $D(x) = \text{Odd}(x)$, we note that

$$\prod_{k \geq 1} \left(1 + x^k\right) \tag{4.8}$$

$$= \prod_{k \geq 1} \frac{\left(1 - x^{2k}\right)}{\left(1 - x^k\right)} \tag{4.9}$$

$$= \frac{\left(1 - x^2\right)}{\left(1 - x^1\right)} \cdot \frac{\left(1 - x^4\right)}{\left(1 - x^2\right)} \cdot \frac{\left(1 - x^6\right)}{\left(1 - x^3\right)} \cdot \frac{\left(1 - x^8\right)}{\left(1 - x^4\right)} \cdots$$

$$= \frac{1}{\left(1 - x^1\right)} \cdot \frac{1}{\left(1 - x^3\right)} \cdot \frac{1}{\left(1 - x^5\right)} \cdots \tag{4.10}$$

$$= \prod_{k \geq 1, \ k \text{ odd}} \frac{1}{\left(1 - x^k\right)}, \tag{4.11}$$

with the equality (4.10) following after the cancellation of all $\left(1 - x^k\right)$ terms, with k even.

Thus, it follows from (4.5), (4.7), (4.9), and (4.11) that $D(x) = \text{Odd}(x)$, and therefore that $p_d(n) = p_{\text{odd}}(n)$, for every nonnegative integer n. In particular, $p_d(2020) = p_{\text{odd}}(2020)$, that is, the number of ways of writing 2020 as a sum of distinct positive integers is equal to the number of ways of writing 2020 as a sum of positive odd integers.
□

Question 13. Suppose that $n \geq 2$ and that $\{x_1, x_2, \ldots, x_n\}$ and $\{y_1, y_2, \ldots, y_n\}$ are two different sets. If the set of all pairwise sums $x_i + x_j$ is equal to the set of all pairwise sums $y_i + y_j$, i.e.,

$$\{x_i + x_j, 1 \leq i \neq j \leq n\} = \{y_i + y_j, 1 \leq i \neq j \leq n\},$$

show that n is a power of 2.

Answer: Assign the polynomials

$$A(t) = \sum_{i=1}^{n} t^{x_i} \quad \text{and} \quad B(t) = \sum_{i=1}^{n} t^{y_i} \tag{4.12}$$

to the sets $\{x_1, x_2, \ldots, x_n\}$ and $\{y_1, y_2, \ldots, y_n\}$, respectively. Then,

$$\begin{aligned}(A(t))^2 - (B(t))^2 &= \left(\sum_{i=1}^n t^{x_i}\right)^2 - \left(\sum_{i=1}^n t^{y_i}\right)^2 \\ &= \sum_{i=1}^n t^{2x_i} + \sum_{1 \leq i \neq j \leq n} t^{x_i} t^{x_j} \\ &\quad - \sum_{i=1}^n t^{2y_i} - \sum_{1 \leq i \neq j \leq n} t^{y_i} t^{y_j}.\end{aligned}$$

The set of all pairwise sums $x_i + x_j$ is equal to the set of all pairwise sums $y_i + y_j$.

$$\{x_i + x_j, 1 \leq i \neq j \leq n\} = \{y_i + y_j, 1 \leq i \neq j \leq n\},$$

Therefore, the number $(A(t))^2 - (B(t))^2$ satisfies

$$\begin{aligned}(A(t))^2 - (B(t))^2 &= \sum_{i=1}^n t^{2x_i} + \sum_{1 \leq i \neq j \leq n} t^{x_i + x_j} \\ &\quad - \sum_{i=1}^n t^{2y_i} - \sum_{1 \leq i \neq j \leq n} t^{y_i + y_j} \\ &= \sum_{i=1}^n t^{2x_i} - \sum_{i=1}^n t^{2y_i}, \quad (4.13)\end{aligned}$$

By rewriting (4.13), it follows that

$$\begin{aligned}(A(t))^2 - (B(t))^2 &= \sum_{i=1}^n (t^2)^{x_i} - \sum_{i=1}^n (t^2)^{y_i} \\ &= A\left(t^2\right) - B\left(t^2\right). \quad (4.14)\end{aligned}$$

Next, consider the polynomial $F(t) = A(t) - B(t)$. Since $A(1) = B(1) = n$, see (4.12), it follows that $F(1) = 0$ and therefore that 1 is a root of $F(t)$. Suppose that 1 is a root

4.1. COUNTING CHALLENGES

with multiplicity k, $k \geq 1$. Then, $F(t) = (t-1)^k H(t)$ for some polynomial $H(t)$.

By using (4.14), we obtain that

$$A(t) + B(t) = \frac{(A(t))^2 - (B(t))^2}{A(t) - B(t)}$$
$$\stackrel{(4.14)}{=} \frac{A(t^2) - B(t^2)}{A(t) - B(t)}.$$

The last fraction can be written in terms of the previously introduced polynomial F.

$$A(t) + B(t) = \frac{F(t^2)}{F(t)}. \qquad (4.15)$$

Moreover,

$$\frac{F(t^2)}{F(t)} = \frac{(t^2-1)^k H(t^2)}{(t-1)^k H(t)}$$
$$= (t+1)^k \frac{H(t^2)}{H(t)}. \qquad (4.16)$$

From (4.15) and (4.16), we find that

$$A(t) + B(t) = (t+1)^k \frac{H(t^2)}{H(t)}. \qquad (4.17)$$

By letting $t = 1$ into (4.17) and using the fact that $A(1) = B(1) = n$, see (4.12), we obtain that

$$2n = A(1) + B(1) = (1+1)^k \frac{H(1)}{H(1)} = 2^k.$$

Therefore, $n = 2^{k-1}$, and we conclude that n is a power of 2. □

4.2 Games, Invariants, Graphs, and Algorithmic Puzzles

Question 1. A transformation of a polygon consists of cutting the polygon into two pieces by a straight line. Then, one of the pieces must be turned over and glued back to the other piece along the edge created by the cut.

Is it possible to start with a square and after finitely many transformations end up with a triangle?

Answer: Observe that after each transformation the area and the perimeter of the polygon do not change. We will now prove that there is no triangle that has the same perimeter and the same area as a square. We will prove that the area of the triangle with sides a, b, and c is strictly smaller than the area of the square whose perimeter is $a + b + c$.

The side of the square is $\frac{a+b+c}{4} = \frac{s}{2}$, where $s = \frac{a+b+c}{2}$ is the semi-perimeter of the triangle. According to Heron's formula, the area of the triangle is

$$A = \sqrt{s(s-a)(s-b)(s-c)}.$$

Using the inequality between the arithmetic and geometric means, we obtain that

$$\begin{aligned} \sqrt{A} &= \sqrt[4]{s(s-a)(s-b)(s-c)} \\ &< \frac{s + (s-a) + (s-b) + (s-c)}{4} \quad (4.18) \\ &= \frac{4s - 2s}{4} \\ &= \frac{s}{2}. \end{aligned}$$

Note that the inequality is always strict because the numbers s, $s-a$, $s-b$, and $s-c$ cannot be all equal and therefore the means inequality from (4.18) must be strict.

4.2. GAMES AND ALGORITHMIC PUZZLES   233

By taking squares of both sides of

$$\sqrt{A} \;<\; \frac{s}{2},$$

we obtain that

$$A \;<\; \frac{s^2}{4} \;=\; \left(\frac{a+b+c}{4}\right)^2.$$

In other words, the area of the triangle is strictly smaller than the area of the square of size $\frac{s}{2}$. Thus, it is not possible to turn a square into a triangle. □

Question 2. In the subtraction of one 3-digit number from another, Alice and Bob fill in the six digits in the following fashion: Alice chooses a number from 0 to 9 and Bob chooses where to enter it as a digit. They continue until all the blank digits are filled. Some numbers may appear more than once, leading zeroes are permitted as well. Alice is trying to maximize the difference, while Bob is trying to minimize it. If both players play their best, what is the difference?

Answer: We will prove that Alice has a strategy that guarantees that the difference is at least 400. Then, we will prove that Bob has a strategy that guarantees that the difference is at most 400. This will imply that if both players play their best, the difference will be 400.

Let us first solve two simpler problems. In the first problem the players are working with one digit numbers α and β in the difference $\alpha - \beta$. Alice can make sure that the difference is at least 4. In her first move she should choose the number 5. If Bob decides to assign this value 5 to the variable α, then Alice should pick 0 as her second number. If Bob assigns $\beta = 5$, then Alice should pick 9 in her second move. Alice will always keep the difference equal to 4. Bob's strategy is the following: If Alice picks a value that is ≥ 5 in her first move, then Bob should assign the value to β. If the number chosen by Alice in her first move is ≤ 4, then Bob should assign that number to the variable α.

Let us now consider the game in which Alice and Bob are working with the difference of two digit numbers

$$\overline{\alpha\beta} - \overline{\gamma\delta}.$$

We will prove that Alice has a strategy that guarantees that the difference is at least 40.

In her first move, Alice should choose the number 5. If Bob places that number as the value for α, then Alice should continue the game by always picking the value 0. If Bob places the number 5 instead of the variable γ, then Alice should continue the game by always choosing the number 9.

We will prove the following Lemma: If Bob makes $\beta = 5$ or $\delta = 5$ in his first move, then Alice has a strategy that guarantees that the difference is at least 40.

If we assume that $\delta = 5$, then Alice should choose 5 as her second number. If Bob places this number 5 instead of β, then the game is reduced to the case of one-digit numbers. Alice can make sure that $\alpha - \gamma \geq 4$, which would imply that

$$\overline{\alpha\beta} - \overline{\gamma\delta} \geq 40.$$

If Bob places this second number 5 instead of α, then Alice should continue the game by always choosing the value 0. The difference will be $\overline{50} - \overline{05} = 45$. If Bob places the second number instead of γ, then Alice should continue the game by always choosing the value 9. The final difference will be $\overline{99} - \overline{55} = 44$.

If we have $\beta = 5$, then Alice should choose 4 in her second move. If Bob puts this number 4 instead of δ, then Alice can use the strategy developed for one digit numbers to

4.2. GAMES AND ALGORITHMIC PUZZLES

achieve $\alpha - \gamma \geq 4$. This would mean that $\overline{\alpha\beta} - \overline{\gamma\delta} \geq 41$. If Bob puts the number 4 instead of α, then Alice should continue the game by always choosing the number 0. The difference will be 45. If Bob puts the number 4 instead of γ, then Alice should continue by always giving the number 9 to Bob. The final difference will be $\overline{95} - \overline{49} = 46$.

We now consider the case in which Alice and Bob play the game with the difference of three digit numbers

$$\overline{abc} - \overline{def}.$$

Alice should keep choosing the number 5 until Bob places one of them instead of a, b, c, or d.

1° If $a = 5$, then Alice should keep choosing numbers 0 until the end of the game. The difference will be at least $\overline{500} - \overline{055} = 445$.

2° If $d = 5$, then Alice should keep choosing numbers 9 until the end. The difference will be at least $\overline{999} - \overline{599} = 400$.

3° If $b = 5$, then we consider two sub-cases.

    3.1° The variable e has already received the value 5. In this case we will use the strategy that we developed for two digit numbers. According to this strategy, Alice can make sure that $\overline{ac} - \overline{df} \geq 40$. Let us now prove that the inequality $\overline{ac} - \overline{df} \geq 40$ implies

$$\overline{a5c} - \overline{d5f} \geq 400.$$

The inequality $\overline{ac} - \overline{df} \geq 40$ can be written as

$$\begin{aligned} \overline{ac} - \overline{df} &= 10a + c - 10d - f \\ &= 10(a - d) + (c - f) \\ &\geq 40. \end{aligned} \quad (4.19)$$

If $c \geq f$, then the inequality $\overline{a5c} - \overline{d5f} \geq 400$ is obvious. If, $c < f$, then from (4.19), we conclude that $a \geq d+5$. The difference $\overline{a5c} - \overline{d5f}$ satisfies

$$\begin{aligned}\overline{a5c} - \overline{d5f} &= 100(a-d) + (c-f) \\ &\geq 500 - 9 \\ &> 400.\end{aligned}$$

3.2° The variable e has not received its value yet. Alice should keep sending the values 4 to Bob until a or d becomes equal to 4. If Bob sets $a = 4$, then Alice should keep sending the numbers 0 until the end of the game. The difference will be at least $\overline{454} - \overline{045} = 409 > 400$. If Bob assigns $d = 4$, then Alice should continue the game by sending the numbers 9. The difference will be at least $\overline{954} - \overline{499} = 455 > 400$.

4° The number c becomes equal to 5 before the numbers a, b, and d. Let us consider the following sub-cases.

4.1° The number f was set to 5 before the number c. In this case, Alice continues with the strategy developed for two digit numbers. Alice can guarantee that $\overline{ab} - \overline{de} \geq 40$, hence $\overline{abc} - \overline{def} \geq 400$.

4.2° The number f has not received its value before the number c. In this case we will consider sub-cases.

4.2.1° The number e has received the value 5 before the number c. Alice should send another number 5. If this additional number 5 is used by Bob to replace the variable b or the variable f, then Alice can use the strategy for the difference of two digit numbers $\overline{\alpha\beta} - \overline{\gamma\delta}$ in which one of β and δ is set to 5.

If Bob assigns this newest number 5 to the variable a, then Alice should continue the

4.2. GAMES AND ALGORITHMIC PUZZLES

game by sending the numbers 0. The difference will be at least $\overline{505} - \overline{050} > 400$.

If Bob assigns this newest number 5 to the variable d, then Alice continues the game by sending the numbers 9. The difference will be at least $\overline{995} - \overline{559} > 400$.

4.2.2° The number e has not received the value 5 before the number c. In this case the number c is the first one to receive the value 5. All other numbers a, b, d, e, and f have not received the values yet. In this case, Alice should send number 4.

If the number 4 gets assigned to the variable f, then Alice continues as in the case 4.1°.

If the number 4 is assigned to the variable e, then Alice continues as in the case 4.2.1°.

If the number 4 is assigned to a, then Alice should keep sending the values 0 until the end. The difference will be 405.

If the number 4 is assigned to d, then Alice should keep sending the values 9 until the end. The difference will be 496.

If the number 4 is assigned to b, then Alice should keep sending the numbers 4 until Bob assigns one of them to either a or d. If $a = 4$, then Alice should keep sending numbers 0 until the end. The difference will be at least $\overline{445} - \overline{044} > 400$. If $d = 4$, then Alice should keep sending numbers 9 until the end. The difference will be at least $\overline{945} - \overline{499} > 400$.

We will now prove that Bob has a strategy that can guarantee that the difference is at most 400.

Note that if Bob manages somehow to put a number greater than or equal to 6 inside the variable d, then Bob will make the difference at most 399. Similarly, the difference will be at most 399 if Alice sends a number ≤ 3 at the time when the variable a is empty.

Bob should keep variable d empty as long as possible to prevent Alice from sending anything ≥ 6. Also, Bob should keep the number a empty to prevent Alice from sending anything ≤ 3.

So, unless Alice wants to make it easy for Bob to get the difference smaller than or equal to 399, her first four numbers must be from $\{4, 5\}$.

When Alice sends the numbers 5, Bob should assign them to e, f, c, b, in this order. When Alice sends the numbers 4, Bob should assign them to b, c, f, e, in this order.

After the first four numbers, Bob has filled the variables b, c, e, and f. When the fifth number arrives, Bob must place it in a or d. If this fifth number is ≥ 5, Bob puts it inside d. The difference $\overline{abc} - \overline{def}$ will be at most

$$\overline{9bc} - \overline{5ef} \leq 400,$$

because Bob has made sure that $b \leq e$ and $c \leq f$.

If the fifth number is smaller than or equal to 4, then Bob puts it inside a. The difference will be at most

$$\overline{abc} - \overline{def} \leq \overline{4bc} - \overline{0ef} \leq 400$$

because $b \leq e$ and $c \leq f$. $\square$

Question 3. Alice and Bob divide a pile of one hundred coins between themselves as follows: Alice chooses a handful of coins from the pile and Bob decides who will get them. This is repeated until all coins have been taken, or until one of them has taken nine handfuls. In the latter case, the other person takes all of the remaining coins in the pile. What is the largest number of coins that Alice can be sure of getting?

4.2. GAMES AND ALGORITHMIC PUZZLES

Answer: The largest number of coins that Alice can be sure of getting is 46. Alice can divide the pile of 100 coins into 16 handfuls of 6 coins each and 1 handful of 4 coins. Since there are 17 handfuls altogether, the game ends as soon as one of the players takes 9 handfuls, consisting of 52 or 54 coins in total. The other player gets the remaining 48 or 46 coins, so Alice can be sure of getting at least 46 coins.

We will now prove the converse: Bob can make sure that Alice gets at most 46 coins. Bob can adopt the strategy of accepting any handful of 6 or more coins and refusing any handful of 5 or fewer coins. If the game ends with Bob accepting 9 handfuls, then Bob gets at least 54 coins and Alice gets at most 46 coins. However, if the game ends otherwise, then Alice gets at most 9 handfuls of at most 5 coins, that is, at most 45 coins in total. Therefore, we again conclude that Alice can be sure of getting at least 46 coins, but not more. $\square$

Question 4. There are four knights on the 3×3 chess board: the two white knights are at the bottom corners, and the two black knights are the two upper corners of the board. The goal is to switch the knights in the minimum number of moves so that the black knights are on the main diagonal and the white knights are on the other diagonal. How do you do it?

Answer: We will prove that it is not possible to move knights in such a way that the black knights are on one diagonal and the white knights on the other.

Let us label the knights with A, B, C, and D. The knights A and B are white and the knights C and D are black. Notice that no knight can occupy the center square. Let us label the remaining squares with $1, 2, \ldots, 8$ as shown in the picture below.

```
    7    6    5
    ┌────┬────┐
    │ D  │ C  │
  8 ├────┼────┤ 4
    │ A  │ B  │
    └────┴────┘
    1    2    3
```

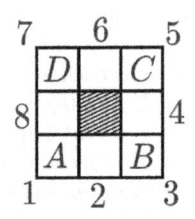

Let us form the graph whose vertices are labeled with 1, 2, ..., 8 and whose edges are drawn between the vertices i and j if and only if a knight can jump from i to j. The graph is the cycle shown in the picture below.

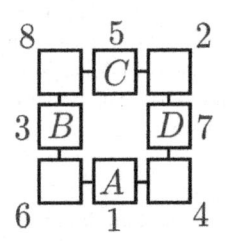

```
     8    5     2
     ┌───┬─C─┬───┐
   3 │ B │   │ D │ 7
     └───┴─A─┴───┘
     6    1     4
```

The game is much easier to analyze by looking at this graph representation. The knights cannot jump over each other in this cycle. Therefore, their order in the graph must always stay the same. The initial configuration has the knights A, B, C, and D in vertices 1, 3, 5, and 7. This means that the order of the knights always must be A, B, C, D in clockwise orientation. In the final desired configuration, the cyclic order of the knights would be different from A, B, C, and D, hence this desired configuration cannot be attained. □

Question 5. Two players start with the sequence 1, 2, ..., 101. They alternate the moves and in each of the

4.2. GAMES AND ALGORITHMIC PUZZLES

moves a player chooses 9 of the numbers and removes them from the sequence. The game is over when only two numbers remain. The player that starts wins $x-54$ dollars from the player that plays second; Here x is the difference between the remaining two numbers after 9 numbers were erased 11 times. Would Bob rather be the first or the second player?

Answer: We will prove that the second player can make sure that $x \leq 55$ and we will prove that the first player can make sure that $x \geq 55$. Hence, if both player plays their best, the difference x of the last two numbers will be 55. Therefore, since the second player sends $x - 54$ dollars to the first player, it is advantageous to choose to be the first player.

It is easy to prove that the second player can guarantee that $x \leq 55$. The second player has to make sure that all of the numbers 57, 58, ..., 101 (which is a total of 45 numbers) are taken. The second player can do that in his 5 moves even if the first player avoids these high numbers. There is a total of 45 numbers that the second player wants to remove. In each of the five moves, the second player can take 9 numbers. The maximal difference among the remaining numbers is $56 - 1 = 55$.

We will now prove that the first player can make sure that $x \geq 55$. In the first move, the first player should take out the numbers 47, 48, ..., 55. After this move, the first player splits the remaining 92 numbers into 46 two-element sets: $\{1, 56\}, \{2, 57\}, \ldots, \{46, 101\}$. We will call these sets *important sets*. The difference of the numbers in each of the important sets is 55. The first player needs to make sure that in the end of the game, the two remaining numbers belong to the same important set. This is easy to achieve. After each move of the second player, the first player has to make a matching move whose goal is the removal of exactly 9 important sets. To accomplish this goal, the first player should follow and adapt to the moves of the opponent. Every time the second player takes out 9 numbers, the first player identifies the important sets from which the numbers were taken. If the second player took z of the important sets in their entirety ($z \in \{0, 1, \ldots, 4\}$), then the second player had to leave $9 - 2z$ important sets with only one element. The first player takes the remaining elements, for a total of $9 - 2z$ numbers. The first player then chooses z of the important sets that are still available and takes all of $2z$ numbers from them. This strategy will guarantee that the last two elements would belong to the same important set and, thus, have their difference equal to 55. $\square$

Question 6. In a computer game, a spy is located on a one-dimensional line. At time 0, the spy is at location a. With each time interval, the spy moves b units to the right if $b > 0$, and $|b|$ units to the left if $b < 0$. Both a and b are fixed integers, but they are unknown to you. Your goal is to identify the location of the spy by asking at each time interval (starting at time 0) whether the spy is currently at some location of your choosing. Devise an algorithm that will find the spy after a finite number of questions.

Answer: The set $\mathbb{Z}^2$ of all ordered pairs (a, b) with $a, b \in \mathbb{Z}$ is countable. Therefore, there exists a sequence (a_0, b_0), (a_1, b_1), (a_2, b_2), ... that contains the entire set $\mathbb{Z}^2$. The two unknown numbers a and b form the unknown pair (a, b) that must belong to this sequence.

The following strategy that will catch the spy: For each $t \in \{0, 1, \ldots\}$ at time t, search for the spy at location $a_t + t b_t$.

Let us denote by k the index for which $a_k = a$ and $b_k = b$. It is obvious that the spy will be caught at the time $t = k$. $\square$

4.2. GAMES AND ALGORITHMIC PUZZLES 243

Question 7. An infection spreads among the squares of an $n \times n$ checkerboard in the following manner: If a square has two or more infected neighbors, then it becomes infected itself. (For example, if one begins with all n infected squares on the diagonal, it spreads to the whole board eventually.) Can you infect the whole board if you begin with fewer than n infected squares?

Answer: No! The key observation is that the total perimeter, that is, the length of the boundary, of all the infected areas (as measured in terms of sides of squares) never increases. Whenever a new square gets infected, the perimeter of the entire infected region either decreases or stays the same. Indeed, for a new square to get infected, it needs to share a side with two or more infected neighboring squares. These two (or more) sides shared with the neighboring squares disappear from the boundary of the infected area. However, our newly infected square can have at most two remaining sides that will be added to the boundary. Hence, the perimeter of the infected area does not increase. It must decrease or stay the same.

Assuming one starts with fewer than n infected squares, the initial perimeter of the infected area is at most $4(n-1)$. On the other hand, when the whole board is infected, the perimeter is $4n$, which is clearly greater than $4(n-1)$. Since the perimeter cannot increase as the squares get infected, it is impossible for the whole board to get infected if one begins with fewer than n infected squares.
□

Question 8. You have to make n pancakes using a skillet that can hold only two pancakes at a time. Each pancake has to be fried on both sides; frying one side of a pancake takes 1 minute, regardless of whether one or two pancakes are fried at the same time. Design an algorithm to do this job in the minimum amount of time. What is the minimum amount of time required?

Answer: If $n = 1$, then we need two minutes to finish frying. We will now prove that for $n \geq 2$, the frying can be done in n minutes. There are $2n$ sides that need to be fried. At most two sides can be fried in a minute. Therefore, at least n minutes are needed for frying. If n is an even number, then can split all pancakes in pairs. The two pancakes in each pair are fried together in exactly 2 minutes. Since there are $\frac{n}{2}$ pairs, the frying can be completed in n minutes.

Assume now that $n \geq 3$ is odd. We will split all the pancakes in two groups: The first group consists of three of the pancakes that we will call A, B, and C. The second group consists of the remaining $n - 3$ pancakes. The number of pancakes in the second group $(n - 3)$ is even, and we have seen how to efficiently fry an even number of pancakes. In other words, we can fry the $n - 3$ pancakes in $n - 3$ minutes. We will now prove that the three pancakes A, B, and C can be fried in three minutes. Let us label the sides of the pancakes by A_1, A_2, B_1, B_2, C_1, and C_2. In the first minute we fry A_1 and B_1. In the second minute, we fry A_2 and C_1. In the third minute, we fry B_2 and C_2. Altogether, the n pancakes will be fried in $(n - 3) + 3 = n$ minutes, even if n is odd!

This completes the proof that for $n \geq 2$ we can finish the frying in n minutes. □

Question 9. Three spiders are trying to catch an ant. All are constrained to the edges of a cube. Each spider can move at least one third as fast as the ant can. Can the spiders always catch the ant?

Answer: We will prove that there exists a strategy that spiders can use that will guarantee them to catch the ant.

4.2. GAMES AND ALGORITHMIC PUZZLES

Let A_0, B_0, C_0, and D_0 be the vertices of the cube on the bottom face in counter-clockwise orientation. Let A_1, B_1, C_1, and D_1 be the corresponding vertices on the top face. Denote by A_H and A_L the points on the edge A_1A_0 that divide A_1A_0 in three equal parts. The point A_H is the higher one (closer to A_1), while A_L is the lower one. More precisely, we have

$$A_1A_H = A_HA_L = A_LA_0 = \frac{1}{3}A_1A_0.$$

In a similar way denote by C_H and C_L the points on the edge C_1C_0 such that

$$C_1C_H = C_HC_L = C_LC_0 = \frac{1}{3}C_1C_0.$$

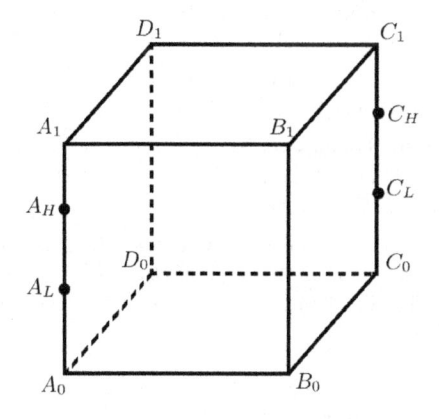

Let us call the spiders α, β, and γ. We will design a strategy for the spiders that consists of four steps as follows:

Step 1. The spiders α and γ should go to the points A_H and C_L, respectively. Once the spiders α and γ reach the points A_H and C_L, Step 1 is over.

Step 2. The spiders α and γ stay in A_H and C_L and the spider β starts moving towards the ant until the ant is forced to enter one of the vertices B_0, B_1, D_0, D_1.

Zones of the cube. In order to prepare for Step 3, we will define the concept of zones. First, we are allowed to assume that the spiders α and γ are in A_H and C_L and that the ant is in one of the vertices B_0, B_1, D_0, D_1. Without loss of generality, assume that the ant is in the vertex B_1.

We will say that the spider α is in the same zone as the ant if one of the following conditions is satisfied:

1° The ant belongs to $B_1 C_1 \cup D_1 C_1$ and the spider α is at the point A_H.

2° The ant belongs to $B_1 A_1 \cup D_1 A_1$ and is at the distance k from A_1; and the spider α is on $A_H A_1$ at the distance $\frac{k}{3}$ from A_1.

3° The ant belongs to $B_0 C_0 \cup D_0 C_0$ and the spider α is at the point A_L.

4° The ant belongs to $B_0 A_0 \cup D_0 A_0$ and is at the distance k from A_0; and the spider α is on $A_L A_0$ at the distance $\frac{k}{3}$ from A_0.

5° The ant is on one of the edges $D_0 D_1$ or $B_0 B_1$ at distance k from the base $A_0 B_0 C_0 D_0$ and the spider α is in the segment $A_L A_H$ at distance $\frac{k}{3}$ from the point A_L.

In an analogous way we define the conditions for the spider γ to be in the same zone as the ant.

Note that if a spider is in the same zone as the ant, the spider can remain in the same zone regardless of how the ant moves. For example, if α is in the same zone as the ant, and the ant starts moving from B_1 to B_0 (or from D_1 to D_0), then α should move from A_H to A_L at the third of the speed of the ant.

4.2. GAMES AND ALGORITHMIC PUZZLES

Moreover, the spider α can make sure that the ant never enters the vertices A_0 or A_1 without getting caught. Assume that the ant starts moving towards A_1. The movement must have started from B_1 or D_1, which means that α is already in A_H. The spider α should start traveling from A_H towards A_1 and reach it at the same time as the ant. If the ant starts moving back, the spider should go back towards A_H. A similar strategy should be used if the ant starts moving towards A_0.

Since we assumed that the ant is in B_1, the spider α is in the same zone as the ant.

Step 3. The goal of step 3 is to have both spiders in the same zone as the ant.

The spider α should use the above strategy to stay in the same zone as the ant and also not allow the ant to enter either of the vertices A_1 or A_0. The spider γ should wait in C_L until the ant enters its zone. The spider β will be the one chasing the ant with the goal of forcing it to move down to B_0 or D_0. Note that the ant does not have a way to escape that destiny. It cannot go to A_1 because the spider α will catch it. Therefore, it does not have a loop on the top face of the cube where it can spend an infinite amount of time running away from β. Thus, the ant should eventually reach B_0 or D_0 and at that point the step 3 is complete.

Now, both spiders α and γ are in the same zone as the ant.

Step 4. The ant now cannot visit the vertices A_0, C_0, A_1, and D_1. The remainder of the cube does not have loops. Therefore, the spider β can now catch the ant. □

Question 10. You are given 8 unit cubes such that 24 of the faces are painted blue and 24 of the faces are painted red. Is it always possible to use these cubes to form a

248   CHAPTER 4. COMBINATORIAL PUZZLES

$2 \times 2 \times 2$ cube that has the same number of blue and red unit squares on its surface?

Answer: At first, arrange the unit cubes into a $2 \times 2 \times 2$ cube arbitrarily. Denote by R the number of visible red faces. Then, the number of visible blue faces is $24 - R$. Assume, without loss of generality, that $R < 12$.

Next, modify the arrangement of cubes by applying the following three rotations to each of the unit cubes. In each of the unit cubes identify the three lines that pass through the centers of opposite faces. Then, make a $90°$ rotation around each of these three lines. After this modification, each of the visible faces becomes invisible and vice versa. The number of the visible red faces goes from R to $24 - R$, and the number of visible blue faces goes from $24 - R$ to R.

Since the number of red faces changes by 1 at each step, in the process of going from R visible red faces to $24 - R$ visible red faces in increments of ± 1, there will be a configuration where the number of visible red faces is exactly 12. For this particular configuration, the number of visible blue faces was also 12.

We conclude that is it always possible to use these cubes to form a $2 \times 2 \times 2$ cube that has the same number, 12, of blue and red unit squares on its surface.   □

Question 11. Jerry the mouse eats his way through a $3 \times 3 \times 3$ cube of cheese by eating all the $1 \times 1 \times 1$ sub–cubes. If Jerry starts at a corner sub–cube and always moves onto an adjacent sub-cube (sharing a face of area 1), can it do this and eat the center sub-cube last? Ignore gravity!

Answer: Let us paint all the unit cubes alternatively in black and white in such a way that two cubes that share a face are of different colors. If one corner cube is black,

4.2. GAMES AND ALGORITHMIC PUZZLES

then all the other corners cubes are also black and the center cube is white. Altogether, there are 14 black and 13 white cubes. The first cube that Jerry eats is black, the second is white, the third is black, and so on. Thus, the last cube that Jerry eats is the 27th cube and must be black. Since the center cube is white, we conclude that the last cube that Jerry eats cannot be the central one. □

Question 12. You want to invert a set of n upright cups by a series of moves in which $n - 1$ cups are turned over at once. Show that this can always be done if n is even and it can never be done if n is odd.

Answer: First, consider the case when n is odd. All the cups being upright at the beginning, the initial configuration of cups has an odd number of upright cups. Since $n - 1$ is even, each move in which $n - 1$ cups are turned over changes the status of an even number of cups, hence preserving the total number of upright cups being odd. Therefore, when n is odd, one cannot invert all the cups, since in that case the number of upright cups would be 0, an even number.

Next, we consider the case when n is even. We present a simple procedure to invert all the cups. For every subset of size $n - 1$ of the n cups, perform the move of turning over all the cups in the subset. The number of moves in this procedure is the number of subsets of size $n-1$, that is, $\binom{n}{n-1} = n$. Moreover, each cup belongs to all but one of these subsets, the exception being the subset of $n - 1$ cups that does not contain the cup itself. Thus, every cup belongs to exactly $n-1$ subsets of size $n-1$, and, hence, it is turned over exactly $n-1$ times. Since n is even, $n-1$ is odd, and therefore each cup is turned over an odd number of times. At the end, each cup will be inverted. □

250 CHAPTER 4. COMBINATORIAL PUZZLES

Question 13. Coins of various sizes are placed on a table, with some touching others. As many times as you wish, you may choose a coin and turn it over along with every other coin that it touches. If all the coins start out showing heads, is it always possible to change them to all tails using these moves?

Answer: Yes, it is always possible to change all the coins to tails using these moves! We will prove this claim by complete induction over n, the number of coins on the table. Note that we will refer to choosing a coin and turning it over along with all the other coins it touches as an *ordinary* move.

To begin the induction, the base case $n = 1$ holds since an ordinary move on the single coin on the table changes it to tails.

Assume that the claim is true for any initial arrangement of fewer than n coins. Consider an arbitrary arrangement of n coins on the table. By using the inductive hypothesis, we know that, for every coin C on the table, there exists a sequence of ordinary moves that changes all coins other than C from heads to tails. If the coin C ends up showing tails as well at the end of this sequence of ordinary moves, then we are done. Otherwise, for *every* coin C on the table, there is a sequence of ordinary moves that changes all coins other than coin C from heads to tails, while leaving coin C showing heads. This sequence of moves will be referred to as a C–*near–perfect* move.

If n is even, perform the C–*near–perfect* move for *every* coin C on the table! Then, each coin on the table changes its face $n - 1$ times, and since $n - 1$ is odd, all coins on the table end up showing tails at the end of these moves.

If n is odd, we first show that there must be a coin on the table that touches an even number of other coins. Define a graph G with vertices at centers of coins and two vertices being adjacent if and only if the corresponding coins are

4.2. GAMES AND ALGORITHMIC PUZZLES

touching. The degree of each vertex v in G is precisely the number of coins that the coin with center at v touches. Summing up the degrees of all vertices in a graph yields twice the number of edges in that graph, in particular, the sum of all the degrees is an even number. Since n is odd, at least one of the degrees of the n vertices in G must be even. In other words, there must be a coin C^* on the table that touches an even number, say, $2k$ other coins.

Finally, perform an *ordinary* move on coin C^* followed by a C–*near*–*perfect* move for every coin C that is not C^* or one of the $2k$ coins that C^* touches. The coin C^* changes its face

$$1 + (n - 2k - 1) = n - 2k$$

times, which is an odd number, since n is odd. Each coin that touches C^* changes its face

$$1 + (n - 2k - 1) = n - 2k$$

times, which is an odd number, since n is odd. Every coin, other than C^* or one of the $2k$ coins that C^* touches, changes its face

$$0 + (n - 2k - 1) - 1 = n - 2k - 2$$

times, which is an odd number, since n is odd.

Thus, every coin on the table changes its face an odd number of times and therefore ends up showing tails. The completes the proof by mathematical induction. $\square$

Question 14. You have a checker in each of the following grid points: $(0,0)$, $(0,1)$, $(1,0)$, $(2,0)$, $(0,2)$, $(1,1)$. You can make a move as follows: if (x,y) is filled and $(x+1,y)$ and $(x,y+1)$ are both empty, then remove checker from (x,y) and put one checker at $(x+1,y)$ and one checker at $(x,y+1)$. Using these moves, can you remove the checkers from all of the six initial positions?

Answer: First, note that any grid point $(x,y) \in \mathbb{Z}^2$ with either x or y (or both) negative can never hold a checker. Next, assign the weight

$$\frac{1}{2^{x+y}}$$

to a grid point (x,y) with $x \geq 0$ and $y \geq 0$. The initial configuration of checkers at the grid points $(0,0)$, $(0,1)$, $(1,0)$, $(2,0)$, $(0,2)$, $(1,1)$ has the total weight of $2\frac{3}{4}$.

Note that a permissible move of removing a checker from the grid point (x,y) and putting one checker at each of the grid points $(x+1,y)$ and $(x,y+1)$ does not change the total weight of the configuration of checkers! Indeed,

$$\frac{1}{2^{x+y}} = \frac{1}{2^{(x+1)+y}} + \frac{1}{2^{x+(y+1)}},$$

so the total weight of the configuration of checkers remains invariant, regardless of the moves played.

We now calculate the total weight of all the grid points $\{(x,y) : x \geq 0, y \geq 0\}$. The total weight of the grid points $\{(x,y) : y \geq 0\}$ for a fixed $x \geq 0$ is

$$\begin{aligned}\sum_{y=0}^{\infty} \frac{1}{2^{x+y}} &= \frac{1}{2^x} \sum_{y=0}^{\infty} \frac{1}{2^y} \\ &= \frac{1}{2^x} \cdot 2 \\ &= \frac{1}{2^{x-1}};\end{aligned}$$

here, we used the geometric series

$$\sum_{y=0}^{\infty} \frac{1}{2^y} = \frac{1}{1-\frac{1}{2}} = 2. \qquad (4.20)$$

Therefore, the total weight of all the grid points $\{(x,y) : x \geq 0, y \geq 0\}$ is

$$\sum_{x=0}^{\infty}\sum_{y=0}^{\infty} \frac{1}{2^{x+y}} = \sum_{x=0}^{\infty} \frac{1}{2^{x-1}} = 2 \cdot \sum_{x=0}^{\infty} \frac{1}{2^x} = 4,$$

4.2. GAMES AND ALGORITHMIC PUZZLES 253

where, for the last equality, we used (4.20) once again.

Since the total weight of $2\frac{3}{4}$ of the initial checkers configuration remains invariant as the permissible moves are played, the rest of the grid points, namely $\{(x,y) : x \geq 0, y \geq 0\} \setminus \{(0,0),(0,1),(1,0),(2,0),(0,2),(1,1)\}$, have a total weight of $4 - 2\frac{3}{4} = 1\frac{1}{4}$ which cannot accommodate checkers with initial weight $2\frac{3}{4}$. In other words, we can never remove checkers from all of the six initial positions.
□

Question 15. There are 100 rest stops in a forest and 1000 trails, each connecting a pair of rest stops. Each trail e has a level of difficulty $d(e)$; no two trails have the same difficulty. An intrepid hiker decided to spend a vacation by hiking 20 trails of ever increasing difficulty. Can he be sure that it can be done? He is free to choose the starting rest stop and the 20 trails form a sequence where the start of one trail is the end of a previous one.

Answer: Yes, he can always find a sequence of 20 trails where the start of one trail is the end of a previous one! Place 100 hikers, one at each rest stop. Then, consider the trails, one by one, in the increasing order of difficulty. When considering trail e, connecting the rest stops s and t, make the hikers currently at those rest stops switch places by hiking on trail e in opposite directions. Let P_v be the sequence of trails traversed by the hiker starting out of rest stop v. Then, by construction, for each of the 100 hikers, the corresponding sequence P_v consists of trails of strictly increasing difficulty. Moreover, note that each trail shows up in exactly two of those sequences. By denoting $\ell(P_v)$ the number of trails in P_v, we obtain that

$$\sum_v \ell(P_v) \;=\; 2 \cdot 1000.$$

Therefore, among the 100 hikers, there must be a hiker, say, starting out of rest stop v, with $\ell(P_v) \geq \frac{2000}{100} = 20$. In other words, our intrepid hiker can start out of rest stop v and follow the sequence P_v of at least 20 trails of ever increasing difficulty. $\square$

Question 16. A room has n computers, less than half of which are damaged. It is possible to query a computer about the status of any computer. A damaged computer could give wrong answers. How can you discover an undamaged computer?

Answer: A computer replies with "damaged" or "undamaged" to a query on the status of another computer. Our algorithm for discovering an undamaged computer in the room is based on the following observation: Assuming that there is a linked list of computers where each computer except the last is queried on the status of its successor in the list, if the replies to all queries are "undamaged" and if there is at least one undamaged computer in the list, then the last computer in the list is definitely undamaged! Our algorithm constructs such a linked list as detailed below.

Start with an empty list. Repeatedly affix one of the remaining computers to the end of the list. If the resulting list has at least two computers and if the penultimate computer replies with "damaged" to the query on the status of the last computer in the list, then remove both of these computers from the list and discard them from the room forever. At least one of the two computers discarded is damaged, so undamaged computers in the room remain in the majority throughout the process of building the linked list. The linked list must be non-empty after all computers have been processed, and there must be at least one undamaged computer in that list. Then, the last com-

4.2. GAMES AND ALGORITHMIC PUZZLES

puter in the list must be undamaged. The total number of queries is exactly $n-1$. □

Question 17. There are $2n$ coins in a bag. Their values are 1, 2, ..., $2n$. Two players, A and B, alternate their moves. A starts the game. In each of the moves, the player takes a coin from the bag and puts in one of the two piles. After all coins are taken out, the player B takes the pile with more money. The player A takes the smaller pile. Let β be the amount of money that B receives. Let α be the money that A receives. What is $\beta - \alpha$ if both players play their best?

Answer: We will prove that $\beta - \alpha = 3n$.

First we will prove that B can guarantee that $\beta - \alpha \geq 3n$. The player B divides the coins into n pairs: one *sticky* pair consisting of coins $(2n-1, 2n)$, and $(n-1)$ *thorny* pairs consisting of coins $(2k-1, 2k)$ for $k \in \{1, \ldots, n-1\}$. The player B needs to make sure that two *sticky* coins end up together in the same pile; while *thorny* coins from each of the pairs end up in opposite piles. The pile with two sticky coins will be the one that B takes in the end. Its value will be at least $2n + 2n - 1 + S$, where S is the sum of all the smaller numbers in each of the thorny pairs. The money that A takes will be at most L, where L is the sum of the larger numbers in each of the thorny pairs. Clearly,

$$\beta - \alpha \geq 4n - 1 + (S - L) \geq 4n - 1 - (n-1) = 3n.$$

Our next step is to prove that A can guarantee that $\beta - \alpha \leq 3n$. The strategy that A needs to use is the *greedy* strategy: In each of the moves, the player A should take the largest coin from the bag and put it in the pile that has the smaller sum.

Although this greedy strategy is intuitive, it is a bit messy to prove that it guarantees $\beta - \alpha \leq 3n$. We can use induction on i to prove the following statement:

$P(i) \equiv$ If A uses the greedy strategy, then

$$\beta_i - \alpha_i \leq 4n - i,$$

where β_i and α_i are the sums of coins in the bigger and the smaller pile after $2i$ moves (i moves of A and i moves of B).

The statement $P(1)$ is obvious. Assume that $P(i)$ is true, and let us prove $P(i+1)$. According to induction hypotesis, we have

$$\beta_i - \alpha_i \leq 4n - i.$$

Since A is playing greedy, A will put the biggest available coin to the pile whose sum is α_i.

Let us denote by α'_{i+1} and β'_{i+1} the sum of the coins in the piles after the $2(i+1)$ moves, if the sums after $2i$ moves were α_i and β_i. There are two cases:

$1°$ $\alpha'_{i+1} \leq \beta'_{i+1}$.

$2°$ $\alpha'_{i+1} > \beta'_{i+1}$.

The case $1°$ corresponds to the situation in which the smaller pile stays smaller and the bigger pile stays bigger, i.e. $\alpha_{i+1} = \alpha'_{i+1}$ and $\beta_{i+1} = \beta'_{i+1}$.

In the case $2°$ the bigger pile turns into a smaller between the moves $2i$ and $2(i+1)$.

The case $1°$ is easier to handle, and case $2°$ is more messy. However, the inequality in case $2°$ is more strict since case $2°$ corresponds to an incompetent play of the opponent B.

Let's do case $1°$ first. Since the biggest coin is used to improve the difference $\alpha'_{i+1} - \alpha_i$, we are sure that $\alpha'_{i+1} -$

4.2. GAMES AND ALGORITHMIC PUZZLES 257

$\alpha_i \geq \beta'_{i+1} - \beta_i + 1$. Therefore,

$$\begin{aligned}
\beta_{i+1} - \alpha_{i+1} &= \beta_i - \alpha_i + (\beta_{i+1} - \beta_i - \alpha_{i+1} + \alpha_i) \\
&\leq 4n - i + (\beta'_{i+1} - \beta_i - \alpha'_{i+1} + \alpha_i) \\
&\leq 4n - i - 1.
\end{aligned}$$

Let us now consider the case 2°. Since A played greedy in each of the first i moves, we are sure that the coins $2n$, $2n - 1$, ..., $2n - i + 1$ are all gone. The biggest available coin has value $2n - i$ at most. The difference $\alpha'_{i+1} - \beta'_{i+1}$ is now positive. The difference $\alpha_i - \beta_i$ was non-positive. We need to prove that $\alpha'_{i+1} - \beta'_{i+1} \leq 4n - i - 1$. The difference $\alpha'_{i+1} - \beta'_{i+1}$ is the biggest if both players A and B contributed the coins $2n - i$ and $2n - (i+1)$ to the pile with the original sum α_i. Therefore

$$\begin{aligned}
\alpha'_{i+1} - \beta'_{i+1} &= \alpha_i - \beta_i + (\alpha'_{i+1} - \alpha_i) - (\beta'_{i+1} - \beta_i) \\
&\leq 0 + (2n - i + 2n - i - 1) - 0 \\
&= 4n - 2i - 1 \\
&< 4n - i - 1. \quad \square
\end{aligned}$$

Chapter 5

Tidbits from Other Areas of Mathematics

5.1 Number Theory

Question 1. Find the smallest positive integer divisible by 225 whose only digits are 0 and 1.

Answer: Note that $225 = 9 \cdot 25$. A number is divisible by 25 if and only if its last two digits are 00, 25, 50, or 75. Since the only digits of our number are 0 and 1, it follows that the last two digits of the number are 00. A number is divisible by 9 if and only if the sum of its digits is divisible by 9. Then, the smallest number divisible by 9 that ends in 00 and whose only digits are 0 and 1 has 9 digits equal to 1 in front of the zeroes. This number is

$$11{,}111{,}111{,}100. \quad \square$$

Question 2. You are given three piles with 5, 49, and 51 pebbles respectively. Two operations are allowed:

(a) merge two piles together, or

(b) divide a pile with an even number of pebbles into two equal piles.

Is there a sequence of operations that would result in 105 piles with one pebble each?

Answer: No! Each of the three piles at the beginning has an odd size. Hence, the first operation must be (a), that is, we must merge two existing piles together, resulting in two piles with possible sizes $\{51, 54\}$, $\{5, 100\}$, or $\{49, 56\}$. Hence, after the first operation, we end up with an even-sized pile and an odd-sized pile.

Denote by d the greatest common divisor of these two pile sizes. In particular, $d = 3$ if the piles are $\{51, 54\}$; $d = 5$ if the piles are $\{5, 100\}$; and $d = 7$ if the piles are $\{49, 56\}$.

Note that, since d is odd and greater than 1, no matter what sequence of operations is carried out (merging two existing piles or dividing an even–sized pile into two equal piles), the size of every pile size will continue to be a multiple of d. Since $d > 1$, we can never reach a configuration where all piles are of size 1.   □

Question 3. Find the smallest positive value of $33^m - 7^n$, where m and n are positive integers.

Answer: For $m = n = 1$, we find that $33^m - 7^n = 26$. Considering the remainder when $33^m - 7^n$ is divided by certain numbers, we will show that no positive value smaller than 26 is attainable! For convenience, we will use the standard notation $a \equiv b \pmod{m}$ which means that when $(a - b)$ is divisible by m.

Consider the remainder of $33^m - 7^n$ when divided by 16. We have $33 \equiv 1 \pmod{16}$, hence $33^m \equiv 1 \pmod{16}$, for all positive integers m. Since $7^2 \equiv 1 \pmod{16}$, we deduce that $7^n \equiv 7 \pmod{16}$ if n is odd, and $7^n \equiv 1 \pmod{16}$ otherwise. Therefore, for all positive integers m and n,

5.1. NUMBER THEORY

either

$$33^m - 7^n \equiv 0 \pmod{16}, \text{ or}$$
$$33^m - 7^n \equiv -6 \equiv 10 \pmod{16}.$$

Hence, the only possible positive values of $33^m - 7^n$ smaller than 26 are 10 and 16.

Next, consider the remainder of $33^m - 7^n$ when divided by 3. Since 33 is divisible by 3, it follows that $33^m \equiv 0 \pmod{3}$, for all positive integers m. Since $7 \equiv 1 \pmod{3}$, we have that $7^n \equiv 1 \pmod{3}$ for all positive integers n. Therefore, for all positive integers m and n, $33^m - 7^n \equiv -1 \equiv 2 \pmod{3}$. Notice that $10 \equiv 1 \pmod{3}$ and $16 \equiv 1 \pmod{3}$. Therefore, the values 10 and 16 cannot be expressed as $33^m - 7^n$.

We conclude that 26 is the smallest positive value of $33^m - 7^n$, where m and n are positive integers. $\square$

Question 4. Given any set of seventeen integers, show that there is at least one subset of nine integers whose sum is divisible by 9.

Answer: We first show that every set of 5 integers has a subset of size three whose elements have the sum divisible by 3. Assume the contrary: There exists a set of 5 integers

$$S = \{a_1, a_2, a_3, a_4, a_5\}$$

such that no 3-element subset has a sum divisible by 3. This implies that no three elements of S have the same remainder when divided by 3. Therefore, there exists at least one element of S that gives remainder 0, at least one element that gives the remainder 1, and at least one element that gives the remainder 2. However, the sum of these three elements is divisible by 3. This proves that every set of 5 integers has a subset of size three whose elements have a sum divisible by 3.

Consider now an arbitrary set T with 17 elements. Since $17 > 5$, there exists a subset

$$T_1 = \{x_{11}, x_{12}, x_{13}\}$$

with three elements whose sum is divisible by 3. Then, the set $T \setminus T_1$ contains $14 > 5$ elements, and therefore it contains a subset

$$T_2 = \{x_{21}, x_{22}, x_{23}\}$$

with three elements whose sum is divisible by 3. Continuing this way, we can find subsets T_3, T_4, and T_5 such that each subset has three elements whose sum is divisible by 3.

For each $i \in \{1, 2, \ldots, 5\}$, denote by τ_i the sum of elements of the set T_i. The numbers τ_i are divisible by 3. Then, the set

$$Q = \left\{\frac{\tau_1}{3}, \frac{\tau_2}{3}, \frac{\tau_3}{3}, \frac{\tau_4}{3}, \frac{\tau_5}{3}\right\}$$

consists of 5 integers and therefore it contains a subset of size 3 whose elements have a sum divisible by 3. Let k, l, m be the three indices for which

$$3 \mid \left(\frac{\tau_k}{3} + \frac{\tau_l}{3} + \frac{\tau_m}{3}\right).$$

Then,

$$9 \mid (\tau_k + \tau_l + \tau_m).$$

Since $\tau_k + \tau_l + \tau_m$ is the sum of the 9 elements from the set $T_k \cup T_l \cup T_m$, we conclude that we found a subset with 9 integers of the initial set of 17 integers whose sum is divisible by 9. $\square$

Question 5. The integers greater than zero are painted such that every number is either red or blue. Both paints are used; blue number + red number = blue number, and blue number × red number = red number.

5.1. NUMBER THEORY

Given only this information, for each of the following decide whether it is a blue number, a red number, or could be either:

red number × red number?

red number + red number?

blue number × blue number?

blue number + blue number?

Answer: First, we prove that 1 has to be blue. Indeed, if 1 were red, then let $n \geq 2$ be any blue positive integer, which must exist since both paints are used. Since $n \times 1 = n$ and "blue number × red number = red number," it would follow that n is red, which is a contradiction. Therefore, 1 is blue.

Let $m \geq 2$ denote the smallest red positive integer (which must exist since both paints are used). Note that the integers $1, 2, \ldots, m-1$ are all blue. Every positive non-multiple of m can be represented as $a \times m + b$, for $a, b \in \mathbb{Z}$, $a \geq 0$, and $1 \leq b \leq m-1$. Repeatedly using "blue number + red number = blue number", we deduce that every non-multiple of m is blue.

Next, we turn our attention to the color of the multiples $a \times m$, $a \in \mathbb{Z}$, of m. If a is a non-multiple of m, using "blue number × red number = red number" we deduce that $a \times m$ is red. If a itself is a multiple of m, then $a+1$ is a non-multiple of m, and hence blue. Using "blue number × red number = red number", we conclude that $(a+1) \times m$ is red. If $a \times m$ were blue, then "blue number + red number = blue number" would imply that $(a+1) \times m = a \times m + m$ is blue as well, which is a contradiction. Hence, $a \times m$ must be red.

We conclude that, if $m \geq 2$ denotes the smallest red positive integer, then all the multiples of m are red integers and all the numbers that are not multiples of m are blue integers.

We can now decide the outcomes of the four expressions:

- "red number × red number = red number,"
  since the product of two multiples of m is also a multiple of m.
- "red number + red number = red number,"
  since the sum of two multiples of m is also a multiple of m.
- "blue number × blue number can be either red or blue,"
  For example, if $m = 6$, then 2, 3, 4 are blue, and 6 is red. Thus, $2 \times 3 = 6$ is red, while $2 \times 2 = 4$ is blue.
- "blue number + blue number can be either red or blue,"
  Similarly, if $m = 6$, then 2, 3, 4, 5 are blue, and 6 is red. However, $2 + 3 = 5$ is blue and $2 + 4 = 6$ is red. $\square$

Question 6. Find all integer solutions of
$$x^3 + 2y^3 = 4z^3.$$

Answer: The only solution to this equation is
$$(x, y, z) = (0, 0, 0).$$

To see this, let (x, y, z) be a solution to the equation
$$x^3 + 2y^3 = 4z^3. \tag{5.1}$$

Since
$$x^3 = 4z^3 - 2y^3, \tag{5.2}$$

it follows that x is an even number and therefore $x = 2x_1$, where x_1 is an integer. Then, (5.2) becomes
$$8x_1^3 = 4z^3 - 2y^3$$

5.1. NUMBER THEORY

and therefore
$$y^3 = 2z^3 - 4x_1^3. \tag{5.3}$$

Thus, y is an even number and therefore $y = 2y_1$, where y_1 is an integer. Then, (5.3) becomes
$$8y_1^3 = 2z^3 - 4x_1^3$$

and therefore
$$z^3 = 2x_1^3 + 4y_1^3, \tag{5.4}$$

Thus, z is an even number and therefore $z = 2z_1$, where z_1 is an integer. Then, (5.4) becomes
$$8z_1^3 = 2x_1^3 + 4y_1^3$$

and therefore
$$x_1^3 + 2y_1^3 = 4z_1^3,$$

which is the same equation as (5.1).

In other words, we showed that, if (x, y, z) is a solution to the equation (5.1) in $\mathbb{Z}^3$, then $(x_1, y_1, z_1) = \left(\frac{x}{2}, \frac{y}{2}, \frac{z}{2}\right)$ is also a solution to (5.1) in $\mathbb{Z}^3$.

Assume that nonzero solutions to the equation (5.1) exist. Let $(x, y, z) \neq (0, 0, 0)$ be the nonzero solution (or one of the solutions) for which $x^2 + y^2 + z^2$ is minimal. Such triple (x, y, z) must exist because we are working with integer solutions only. Then, $(x_1, y_1, z_1) = \left(\frac{x}{2}, \frac{y}{2}, \frac{z}{2}\right)$ is also an integer solution to (5.1) and
$$x_1^2 + y_1^2 + z_1^2 = \frac{1}{4}\left(x^2 + y^2 + z^2\right). \tag{5.5}$$

Since we assumed that $x^2 + y^2 + z^2$ is minimal, it follows from (5.5) that $x^2 + y^2 + z^2 = 0$, which means that $x = y = z = 0$ and contradicts the assumption that $(x, y, z) \neq (0, 0, 0)$.

We conclude that the only solution to the equation (5.1) is $(x, y, z) = (0, 0, 0)$. $\square$

5.2 Geometry

Question 1. You are given a finite number of points in the plane with the property that any line that contains two of these points contains at least three of these points. Show that all the points must lie on a straight line.

Answer: We will give a proof by contradiction.

Suppose that the points are not collinear, that is, they do not all lie on a straight line. Denote by P the set of points, and let L be the set of lines they define. Consider the set of pairs of points and lines

$$\mathcal{D} = \{(p, \ell) \in P \times L \mid p \notin \ell\}$$

where, for the pair (p, ℓ), the line ℓ from L does not contain the point p from P.

By the assumption of non-collinearity, the set $\mathcal{D}$ is not empty. For each pair $(p, \ell) \in \mathcal{D}$ record the distance $d(p, \ell)$ between p and ℓ. Let (p^*, ℓ^*) be the pair with the minimum distance (breaking the ties arbitrarily).

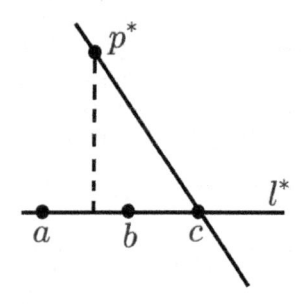

Drop a perpendicular from p^* onto the line ℓ^*. Since every line contains at least three points, it follows that the line ℓ^* contains at least three points of P, say a, b, c, appearing in that order along ℓ^*.

5.2. GEOMETRY

By the pigeonhole principle, two of these points, say b and c, must be on the same side of the intersection of ℓ^* with the perpendicular from p^*. Denote by $\tilde{\ell} \in L$ the line passing through the points p^* and c.

Note that the distance $d(b, \tilde{\ell})$ from b to the line $\tilde{\ell}$ is strictly smaller than $d(p^*, \ell^*)$, the distance from p^* to the line ℓ^*. This is because the triangle p^*bc is obtuse and the altitude from the vertex b of the obtuse angle must be the shorter than the altitude from the point p^*. This contradicts the minimality property of the pair (p^*, ℓ^*).

We conclude that the set $\mathcal{D}$ must be an empty set, and therefore that all the point lie on a straight line. $\quad\square$

Question 2. Find all functions $f : \mathbb{R}^2 \to \mathbb{R}$ defined over the two-dimensional plane with the property that, for any square $ABCD$,

$$f(A) + f(B) + f(C) + f(D) = 0. \qquad (5.6)$$

Answer: Let O be an arbitrary point in the plane, and consider a square $ABCD$ in the plane with center at the point O. From (5.6), it follows that

$$f(A) + f(B) + f(C) + f(D) = 0. \qquad (5.7)$$

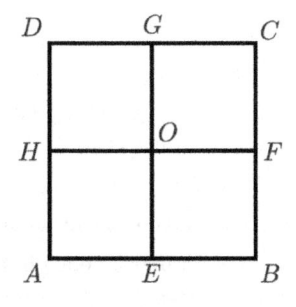

Denote by E, F, G, and H the midpoints of the sides AB, BC, CD, and DA of the square, respectively.

Then, the quadrilaterals $AEOH$, $BEOF$, $CGOF$, and $DGOH$ are also squares, and, from the property (5.6) of the function f, we obtain that

$$f(A) + f(E) + f(O) + f(H) = 0; \quad (5.8)$$
$$f(B) + f(E) + f(O) + f(F) = 0; \quad (5.9)$$
$$f(C) + f(G) + f(O) + f(F) = 0; \quad (5.10)$$
$$f(D) + f(G) + f(O) + f(H) = 0. \quad (5.11)$$

Moreover, $EFGH$ is also a square, and therefore

$$f(E) + f(F) + f(G) + f(H) = 0. \quad (5.12)$$

By adding (5.8–5.11) and using (5.7) and (5.12), we conclude that

$$\begin{aligned} 0 &= f(A) + f(B) + f(C) + f(D) \\ &\quad + 2\left(f(E) + f(F) + f(G) + f(H)\right) \\ &\quad + 4f(O) \\ &= 4f(O). \end{aligned}$$

We conclude that $f(O) = 0$ for any point O in the plane and therefore the only function f satisfying (5.6) is the zero function $f = 0$. $\square$

Question 3. What is the largest possible value of n such that there exist n points in the plane with just two different distances between them?

Answer: The largest possible value of n is 5.

Indeed, it is possible to arrange 5 points in the plane with only two different distances between them: place the 5

5.2. GEOMETRY

points at the vertices of a regular pentagon. The distance between any two points is either the length of a side or the length of a diagonal. Since all sides have the same length, and all diagonals have the same length, there are only two different distances between the 5 points.

Next, we show that it is impossible to arrange 6 points in the plane with only two different distances between them.

Suppose to the contrary that there are six points in the plane with only two different distances ℓ and $\mathcal{L}$ between them. Denote one of the points by P and draw the line segments to the remaining five points. By the pigeonhole principle, at least three of these five line segments have the same length, say ℓ. Consider the three other endpoints of these line segments. If any two of them are at distance ℓ, we have an equilateral triangle of side length ℓ, formed by P and those two points. Otherwise, the three other endpoints of these line segments are at distance $\mathcal{L}$ from each other, and so they form an equilateral triangle of side length $\mathcal{L}$. In either case, three of the six points must form an equilateral triangle!

We can therefore assume that three of the six points form an equilateral triangle. Label these points as A, B, and C, and with proper scaling, we can assume that the equilateral triangle ABC has the side length $\ell = 1$. Label the three remaining points as D, E, and F. Consider any of these points, say D. Since D cannot be at distance $\ell = 1$ from all three of the points A, B, C, there are three cases:

(1) D is at distance $\mathcal{L}$ from exactly one of the points A, B, C, say A, and at distance $\ell = 1$ from exactly two of the other points, that is, the points B and C.

(2) D is at distance $\mathcal{L}$ from exactly two of the points A, B, C, say B and C and at distance $\ell = 1$ from exactly one of the other points, that is, the point A.

(3) D is at distance $\mathcal{L}$ from all three of the points A, B, C.

The cases (1), (2), and (3) correspond to the figures below. Note that two middle figures cover two distinct possibilities for Case (2), as there are two different placements for point D along the bisector of segment BC, so that D is at distance $\ell = 1$ from A.

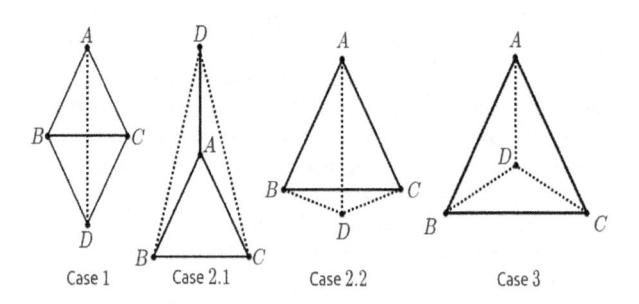

Case 1   Case 2.1   Case 2.2   Case 3

In Case (1), $AD = \mathcal{L} = \sqrt{3}$, as AD is twice the height of the equilateral triangle ABC. In Case (2.1), $BD = CD = \mathcal{L} = \sqrt{1^2 + 1^2 - 2 \cdot 1 \cdot 1 \cdot \cos\left(\frac{5\pi}{6}\right)} = \sqrt{2 + \sqrt{3}}$, using the law of cosines on the triangle ABD, where $\angle BAD = \frac{5\pi}{6}$. In Case (2.2),

$$\begin{aligned} BD = CD = \mathcal{L} &= \sqrt{1^2 + 1^2 - 2 \cdot 1 \cdot 1 \cdot \cos\left(\frac{\pi}{6}\right)} \\ &= \sqrt{2 - \sqrt{3}}. \end{aligned}$$

We used the law of cosines on the triangle ABD, where $\angle BAD = \frac{\pi}{6}$. In Case (3), $AD = BD = CD = \mathcal{L} = \frac{\sqrt{3}}{3}$, as AD is $\frac{2}{3}$ of the height of the equilateral triangle ABC.

Since the values of $\mathcal{L}$ in all of these cases are different, points E and F must fall into precisely the same case as point D! This rules out Case (3) immediately, since there is only one possible location for points D, E, and F. The cases (1), (2.1), and (2.2) correspond to the figures below.

5.2. GEOMETRY

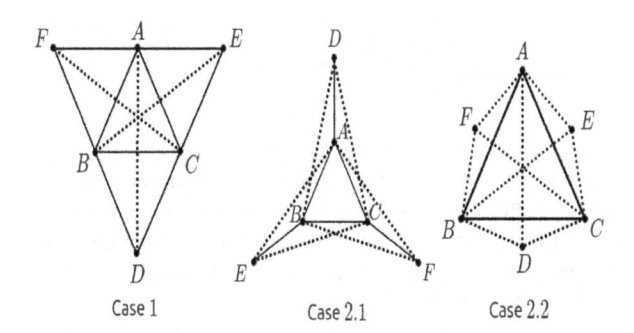

Case 1   Case 2.1   Case 2.2

In Cases (1) and (2.1), the points D and E are clearly at a greater distance than ℓ or $\mathcal{L}$, which is a contradiction.

In Case (2.2),

$$AB = AC = BC$$
$$= AD = BE = CF = \ell = 1;$$
$$BD = DC = CE$$
$$= EA = AF = FB = \mathcal{L} = \sqrt{2 - \sqrt{3}}.$$

Since $\angle BAD = \angle CAD = \frac{\pi}{6}$ and since the triangles ABD and ADC are isosceles, we obtain that

$$\angle BDA = \angle CDA = \frac{5\pi}{12}.$$

Thus, $\angle BDC = \frac{5\pi}{6}$, and, similarly,

$$\angle CEA = \angle AFB = \frac{5\pi}{6}.$$

Since the triangles BDC, CEA and AFB are isosceles, we have $\angle DBC = \angle DCB = \angle ECA = \angle EAC = \angle FAB = \angle FBA = \frac{\pi}{12}$. Therefore, $\angle DCE = \angle EAF = \angle FBD = \frac{\pi}{2}$. Then, from the right triangle DCE, it follows that

$$DE = \sqrt{DC^2 + CE^2} = \mathcal{L}\sqrt{2} = \sqrt{4 - 2\sqrt{3}}$$
$$= \sqrt{\left(\sqrt{3} - 1\right)^2} = \sqrt{3} - 1,$$

a distance different from $\ell = 1$ and $\mathcal{L} = \sqrt{2 - \sqrt{3}}$, which is again a contradiction. □

Question 4. The new campus of University College is a perfect disk of radius 1km. The Coffee Company plans to open 7 coffee shops on campus. Where do they have to be placed in order to minimize the maximum (straight-line) distance that a person anywhere on the campus has to walk to find a coffee shop?

Answer: In order to minimize the maximum distance to a coffee shop, the Coffee Company should place the seven coffee shops at the midpoints of the sides and at the center of the regular hexagon inscribed in the perfect disk of radius 1km, as indicated in the figure below. This way, the maximum distance a person anywhere on campus has to walk to find a coffee shop is exactly 0.5km.

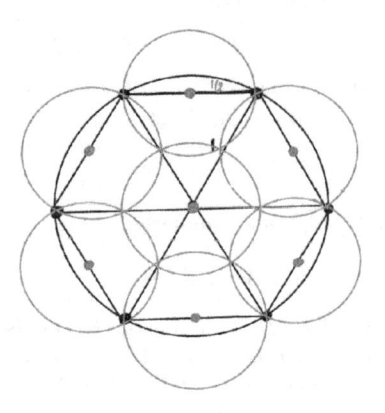

Why is 0.5km optimal?

Suppose that, on the contrary, the Coffee Company could place seven coffee shops on campus so that the the maximum distance from anywhere on campus to a coffee shop would be $r < 0.5$km. Then, the seven disks of radius r, centered at those coffee shops, would cover the entire campus, that is, the entire disk of radius of 1km.

Let us prove that a disk of radius r ($r < 1$) can cover at most $2\sin^{-1}(r)$ of the circumference of the disk of radius 1. Let us denote by A and B the instersections of the two disks. Then $AB \leq 2r$. If O Is the center of the big disk of radius 1 and M the midpoint of AB, then we have

$$AM = \sin \angle AOM.$$

The arc AB has the length equal to $\angle AOB$. This length satisfies

$$\angle AOB = 2\angle AOM = 2\sin^{-1}(AM) \leq 2\sin^{-1}(r).$$

Then, six disks of radius r cover at most $6 \cdot 2\sin^{-1}(r) = 12\sin^{-1}(r)$ of the circumference of the disk of radius 1. Note that, since $r < 0.5$, it follows that $\sin^{-1}(r) < \frac{\pi}{6}$ and therefore

$$12 \cdot \frac{\pi}{6} = 2\pi.$$

Thus, six of the disks of radius r cannot cover the entire circumference of the unit circle, This means that all of the seven disks of radius r must touch the circumference of the disk of radius 1. Since $r < 0.5$, none of those disks would cover the center the disk of radius 1; in other words, a person at the center of the campus would have to walk more than 0.5km to the closest coffee shop, which is a contradiction. □

Question 5. Inscribe a regular N-gon in a circle of radius 1. Draw the $N - 1$ segments connecting a given vertex to the $N - 1$ other vertices. Find the product of the lengths of these segments.

Answer: Place the circle in the complex plane, with the center at the origin and the given vertex of the inscribed regular N–gon located at the point $(1,0)$. The other vertices of the N–gon are located at the remaining N-th roots of unity ξ^k, $k = 1, 2, \ldots, N-1$, where $\xi = e^{2\pi i/N}$. Denote by d_k the length of the segment connecting the vertex at $\xi^0 \equiv (1,0)$ to the vertex at ξ^k. Then, $d_k = |1 - \xi^k|$ and the product Π_N of the lengths of these $N - 1$ segments is

$$\begin{aligned} \Pi_N &= \prod_{k=1}^{N-1} d_k = \prod_{k=1}^{N-1} |1 - \xi^k| \\ &= \left| \prod_{k=1}^{N-1} (1 - \xi^k) \right|. \end{aligned} \qquad (5.13)$$

Next, consider the function

$$f(z) = z^N - 1, \text{ for } z \in \mathbb{C}.$$

The roots of $f(z)$ are $1, \xi, \ldots, \xi^{N-1}$, where $\xi = e^{2\pi i/N}$. Therefore, $f(z)$ can be factored as

$$f(z) = (z-1) \cdot \prod_{k=1}^{N-1} (z - \xi^k). \qquad (5.14)$$

Moreover,

$$f(z) = z^N - 1 = (z-1) \cdot \sum_{k=0}^{N-1} z^k. \qquad (5.15)$$

From (5.14) and (5.15), we obtain that

$$\prod_{k=1}^{N-1} (z - \xi^k) = \sum_{k=0}^{N-1} z^k, \ \forall z \in \mathbb{C}. \qquad (5.16)$$

By letting $z = 1$ in (5.16), we obtain from (5.13) that

$$\Pi_N = \left| \prod_{k=1}^{N-1} (1 - \xi^k) \right| = \left| \sum_{k=0}^{N-1} 1^k \right| = \left| \sum_{k=0}^{N-1} 1 \right| = N.$$

5.2. GEOMETRY

We conclude that the product of the lengths of the $N-1$ segments is equal to N. □

Question 6. Can you draw five rays from the origin in $\mathbb{R}^3$ so that every angle between two of the rays is obtuse?

Answer: No! We give a proof by contradiction. Assume that it is possible to draw five such rays. Denote by $\vec{v_i} = (x_i, y_i, z_i)$, $1 \leq i \leq 5$, the unit vector along the ith ray. Let $\theta_{i,j}$ be the obtuse angle between $\vec{v_i}$ and $\vec{v_j}$. Choose the coordinate system so that $\vec{v_5} = (0, 0, -1)$. Since the angle between $\vec{v_i}$ and $\vec{v_5}$, $i \neq 5$, is obtuse, we conclude that the dot product $\vec{v_i} \cdot \vec{v_5}$ is negative. Thus,

$$\vec{v_i} \cdot \vec{v_5} = \cos\theta_{i,5} < 0. \tag{5.17}$$

Since

$$\vec{v_i} \cdot \vec{v_5} = (x_i, y_i, z_i) \cdot (0, 0, -1) = -z_i,$$

it follows from (5.17) that $-z_i < 0$, and therefore that

$$z_i > 0, \quad \forall\, i = 1:4. \tag{5.18}$$

Next, project the vectors $\vec{v_1}$, $\vec{v_2}$, $\vec{v_3}$, $\vec{v_4}$, onto the xy-plane, and denote by $\vec{w_i} = (x_i, y_i, 0)$ the projection of $\vec{v_i} = (x_i, y_i, z_i)$, for $i = 1:4$. By the pigeonhole principle, out of six possible pairs, there exists at least one pair made of two of the vectors $\vec{w_1}$, $\vec{w_2}$, $\vec{w_3}$, and $\vec{w_4}$ that form an angle that is not obtuse. Without loss of generality, assume that $\vec{w_1}$ and $\vec{w_2}$ are such a pair. Then, the dot product of the vectors $\vec{w_1} = (x_1, y_1, 0)$ and $\vec{w_2} = (x_2, y_2, 0)$ is nonnegative:

$$\vec{w_1} \cdot \vec{w_2} = x_1 x_2 + y_1 y_2 \geq 0.$$

Since $z_1 > 0$ and $z_2 > 0$, see (5.18), it follows that $z_1 z_2 > 0$ and therefore

$$\vec{v_1} \cdot \vec{v_2} = x_1 x_2 + y_1 y_2 + z_1 z_2 > 0.$$

Thus, the vectors $\vec{v_1}$ and $\vec{v_2}$ form an acute angle, which contradicts the assumption that there exist five rays from the origin with obtuse angles between any two rays.

We conclude that it is not possible to draw five rays from the origin in $\mathbb{R}^3$ so that every angle between two of the rays is obtuse. □

5.3 Calculus

Question 1. Given a polynomial f of degree 98 with the property that
$$f(k) = \frac{1}{k}, \ \ \forall \ k \in \{1, 2, \ldots, 99\},$$
find $f(100)$.

Answer: Consider the polynomial
$$g(x) = xf(x) - 1. \tag{5.19}$$

Then, $g(x)$ is a polynomial of degree 99 and $1, 2, \ldots, 99$ are all the roots of $g(x)$. Therefore,
$$g(x) = C\,(x-1)(x-2)\cdots(x-99). \tag{5.20}$$

From (5.19), we find that $g(0) = -1$. By letting $x = 0$ in (5.20), it follows that
$$g(0) = C \cdot (-1)^{99} \cdot 99!.$$

Since $g(0) = -1$, we conclude that $C = \frac{1}{99!}$ and we obtain from (5.20) that
$$g(x) = \frac{1}{99!}\,(x-1)(x-2)\cdots(x-99).$$

For $x = 100$, we find that $g(100) = 1$ and, using (5.19), we conclude that
$$1 = g(100) = 100\,f(100) - 1,$$
and therefore
$$f(100) = \frac{1}{50}. \ \ \square$$

Question 2. How many real solutions does the following equation have:

$$\sqrt[7]{x} - \sqrt[5]{x} = \sqrt[3]{x} - \sqrt{x}?$$

Answer: We first rewrite the equation as

$$\sqrt{x} - \sqrt[3]{x} - \sqrt[5]{x} + \sqrt[7]{x} = 0. \quad (5.21)$$

Note that $x \geq 0$, otherwise $\sqrt{x}$ is not a real number. Assume, therefore, that $x \geq 0$ and define

$$t = x^{\frac{1}{2 \cdot 3 \cdot 5 \cdot 7}}.$$

Then, $x = t^{2 \cdot 3 \cdot 5 \cdot 7}$. From (5.21), it follows that the problem is equivalent to counting the non-negative solutions of the equation

$$t^{105} - t^{70} - t^{42} + t^{30} = 0,$$

which can be written as

$$t^{30} \left(t^{75} - t^{40} - t^{12} + 1 \right) = 0. \quad (5.22)$$

Clearly, one solution of (5.22) is $t = 0$. It remains to count the number of positive roots of the polynomial

$$P(t) = t^{75} - t^{40} - t^{12} + 1. \quad (5.23)$$

Note that

$$\begin{aligned} P'(t) &= 75t^{74} - 40t^{39} - 12t^{11} \\ &= t^{11} \left(75t^{63} - 40t^{28} - 12 \right). \end{aligned} \quad (5.24)$$

We want to find the intervals of $(0, \infty)$ on which P' is positive and the intervals on which it is negative. The sign of the polynomial P' on $(0, \infty)$ is the same as the sign of the polynomial Q defined as

$$Q(t) = 75t^{63} - 40t^{28} - 12. \quad (5.25)$$

5.3. CALCULUS

We write this explicitly as

$$\operatorname{sgn}(P') = \operatorname{sgn}(Q). \tag{5.26}$$

We will find the subintervals of $(0, \infty)$ on which Q is positive and the subintervals on which Q is negative. The derivative of Q satisfies

$$\begin{aligned} Q'(t) &= 75 \cdot 63 \cdot t^{62} - 40 \cdot 28 \cdot t^{27} \\ &= 75 \cdot 63 \cdot t^{27} \left(t^{35} - \frac{32}{135} \right). \end{aligned} \tag{5.27}$$

Since $75 \cdot 63 \cdot t^{27}$ is positive on $(0, \infty)$, the sign of Q' on $(0, \infty)$ is the same as the sign of

$$R(t) = t^{35} - \frac{32}{135}$$

on $(0, \infty)$. In other words,

$$\operatorname{sgn}(Q') = \operatorname{sgn}(R). \tag{5.28}$$

Let

$$\alpha = \left(\frac{32}{135} \right)^{\frac{1}{35}} \tag{5.29}$$

and note that $\alpha \in (0, 1)$. The function $R(t)$ is negative on $(0, \alpha)$ and positive on (α, ∞). From (5.28), it follows that the same is true for Q': the function Q' is negative on $(0, \alpha)$ and positive on (α, ∞). Therefore, the function Q is decreasing on $(0, \alpha)$ and increasing on (α, ∞) and the function Q attains the minimum on $(0, \infty)$ at α.

Recall that $\alpha \in (0, 1)$. From (5.25), we find that $Q(0) = -12$ and $Q(1) = 23 > 0$. Therefore, $Q(\alpha)$ is a negative number. Since $Q(t)$ is a continuous function, $Q(\alpha) < 0$, and $Q(1) > 1$, there exists a real number $\beta \in (\alpha, 1)$ such that $Q(\beta) = 0$. While we do not know exactly what β is, we know that α and β satisfy

$$0 < \alpha < \beta < 1. \tag{5.30}$$

280   CHAPTER 5. MATHEMATICAL TIDBITS

Moreover, the function Q is negative on $(0, \beta)$ and positive on (β, ∞).

From (5.26), it follows that the same is true for P': P' is negative on $(0, \beta)$ and positive on (β, ∞). We conclude that $P(t)$ is decreasing on $(0, \beta)$ and increasing on (β, ∞). Therefore, the minimum of P on $(0, \infty)$ is attained at β.

Note that $P(0) = 1$ and $P(1) = 0$; see (5.23). Since $\beta < 1$ and β is the minimum point of $P(t)$, it follows that $P(\beta) < P(1) = 0$. Thus, the equation $P(t) = 0$ must have one solution $t = \gamma$ for some γ in the interval $(0, \beta)$ and therefore the function P has two zeros on $(0, \infty)$, which are γ and 1.

We conclude that there are three non-negative solutions to the equation (5.22), and therefore that there are three real numbers x that are solutions to the given equation. $\square$

Question 3. Find all the real solutions of the equation

$$\sin(\cos x) = \cos(\sin x).$$

Answer: We will prove that the equation has no solutions in $\mathbb{R}$.

Define the function $g(x)$ as

$$g(x) = \cos(\sin x) - \sin(\cos x).$$

Using the facts that cosine is an even function while sine is an odd function, we find that

$$\begin{aligned}
g(-x) &= \cos(\sin(-x)) - \sin(\cos(-x)) \\
&= \cos(-\sin x) - \sin(\cos x) \\
&= \cos(\sin x) - \sin(\cos x) \\
&= g(x).
\end{aligned}$$

5.3. CALCULUS

Thus, $g(x)$ is an even function. Moreover, since sine and cosine are periodic functions with period 2π, $g(x)$ is periodic with period 2π as well. So, it suffices to show that equation $g(x) = 0$ has no solutions in $[0, \pi]$.

If $\frac{\pi}{2} \leq x \leq \pi$, then $0 \leq \sin x \leq 1 < \frac{\pi}{2}$ and therefore $\cos(\sin x) > 0$. Similarly, if $\frac{\pi}{2} \leq x \leq \pi$, then $-\frac{\pi}{2} < -1 \leq \cos x \leq 0$, and therefore $\sin(\cos x) \leq 0$. Then, $g(x) = \cos(\sin x) - \sin(\cos x) > 0$, and therefore the equation $g(x) = 0$ has no solutions in $\left[\frac{\pi}{2}, \pi\right]$.

Subsequently, we will use the fact that

$$\sin t < t, \ \forall \, t > 0. \tag{5.31}$$

If $0 < x < \frac{\pi}{2}$, then, according to (5.31), $\sin x < x < \frac{\pi}{2}$. Since $\cos x$ is a decreasing function on $\left[0, \frac{\pi}{2}\right]$, it follows that

$$\cos(\sin x) > \cos x. \tag{5.32}$$

Also, if $0 < x < \frac{\pi}{2}$, then $0 < \cos x < 1$. Using the inequality (5.31) for $t = \cos x$, we obtain that

$$\sin(\cos x) < \cos x. \tag{5.33}$$

From (5.32) and (5.33), we find that

$$\cos(\sin x) > \sin(\cos x).$$

Thus, $g(x) = \cos(\sin x) - \sin(\cos x) > 0$, and therefore the equation $g(x) = 0$ has no solutions in $\left(0, \frac{\pi}{2}\right)$.

We conclude that there are no values of x such that $g(x) = 0$, in other words, such that $\sin(\cos x) = \cos(\sin x)$. $\square$

Question 4. Let $f(x)$ be a positive, continuously differentiable function, defined for all real numbers, whose derivative is always negative. Fix any x_0 and run the Newton's method. What is $\lim_{n \to \infty} x_n$?

282   CHAPTER 5. MATHEMATICAL TIDBITS

Answer: The Newton's method recursion is

$$x_{n+1} \;=\; x_n - \frac{f(x_n)}{f'(x_n)}, \quad \forall\, n \geq 0, \qquad (5.34)$$

with x_0 given. If $f(x)$ is a positive function with negative derivative, then $f(x_n) > 0$ and $f'(x_n) < 0$. Thus, $\frac{f(x_n)}{f'(x_n)} < 0$ and it follows from (5.34) that $x_{n+1} > x_n$ for all n. In other words, the sequence $(x_n)_{n\geq 0}$ is strictly increasing and it will either converge to a finite limit or it will go to infinity.

However, the sequence $(x_n)_{n\geq 0}$ cannot be convergent: If there exists a number l such that $l = \lim_{n\to\infty} x_n$, then, by letting $n \to \infty$ in (5.34), we obtain that

$$l \;=\; l - \frac{f(l)}{f'(l)} \quad \Longleftrightarrow \quad f(l) = 0.$$

This is a contradiction with the fact that $f(x)$ is a positive function and therefore that $f(x) > 0$ for any x.

We conclude that the strictly increasing sequence $(x_n)_{n\geq 0}$ goes to infinity, i.e.,

$$\lim_{n\to\infty} x_n \;=\; \infty. \qquad \square$$

Question 5. Find

$$\lim_{n\to\infty} \left| \sin\left(\pi\sqrt{n^2 + n + 1}\right) \right|,$$

where n is a positive integer.

Answer: Note that the function $|\sin(\cdot)|$ is periodic with period π since

$$|\sin(x - \pi)| \;=\; |-\sin(x)| \;=\; |\sin(x)|, \quad \forall\, x \in \mathbb{R}.$$

5.3. CALCULUS

Thus,

$$\lim_{n\to\infty} \left|\sin\left(\pi\sqrt{n^2+n+1}\right)\right|$$
$$= \lim_{n\to\infty} \left|\sin\left(\pi\sqrt{n^2+n+1} - n\pi\right)\right|$$
$$= \lim_{n\to\infty} \left|\sin\left(\pi\left(\sqrt{n^2+n+1} - n\right)\right)\right|. \quad (5.35)$$

Let us introduce A_n as

$$A_n = \sqrt{n^2+n+1} - n. \quad (5.36)$$

Multiplying by the conjugate, we express A_n in a form in which the limit is easy to calculate.

$$\begin{aligned}
A_n &= \frac{\left(\sqrt{n^2+n+1} - n\right)\left(\sqrt{n^2+n+1} + n\right)}{\sqrt{n^2+n+1} + n} \\
&= \frac{(\sqrt{n^2+n+1})^2 - n^2}{\sqrt{n^2+n+1} + n} \\
&= \frac{n+1}{\sqrt{n^2+n+1} + n} \\
&= \frac{1 + \frac{1}{n}}{\sqrt{1 + \frac{1}{n} + \frac{1}{n^2}} + 1}
\end{aligned}$$

and therefore

$$\lim_{n\to\infty} A_n = \lim_{n\to\infty} \left(\sqrt{n^2+n+1} - n\right) = \frac{1}{2}. \quad (5.37)$$

Since the function $|\sin(\cdot)|$ is continuous, we conclude from (5.35) and (5.37) that

$$\begin{aligned}
&\lim_{n\to\infty} \left|\sin\left(\pi\sqrt{n^2+n+1}\right)\right| \\
&= \lim_{n\to\infty} \left|\sin\left(\pi(\sqrt{n^2+n+1} - n)\right)\right| \\
&= \left|\sin\left(\pi \lim_{n\to\infty}(\sqrt{n^2+n+1} - n)\right)\right| \\
&= \left|\sin\left(\frac{\pi}{2}\right)\right| \\
&= 1. \quad \square
\end{aligned}$$

Question 6. Find the 100th derivative of the function
$$\frac{x^2+1}{x^3-x}.$$

Answer: Since $x^3 - x = x(x-1)(x+1)$, we can use partial fractions to find that
$$\frac{x^2+1}{x^3-x} = \frac{1}{x-1} + \frac{1}{x+1} - \frac{1}{x}.$$

Therefore,

$$\begin{aligned}
&\frac{d^{100}}{dx^{100}}\left(\frac{x^2+1}{x^3-x}\right) \\
&= \frac{(-1)^{100} \cdot 100!}{(x-1)^{101}} + \frac{(-1)^{100} \cdot 100!}{(x+1)^{101}} \\
&\quad - \frac{(-1)^{100} \cdot 100!}{x^{101}} \\
&= \frac{100!}{(x-1)^{101}} + \frac{100!}{(x+1)^{101}} - \frac{100!}{x^{101}}. \quad \square
\end{aligned}$$

Question 7. Compute the integral
$$\int_{-1}^{1} \frac{\sqrt[3]{x}}{\sqrt[3]{1-x}+\sqrt[3]{1+x}}\, dx.$$

Answer: Let
$$f(x) = \frac{\sqrt[3]{x}}{\sqrt[3]{1-x}+\sqrt[3]{1+x}}.$$

Note that the function $f(x)$ is an odd function:
$$\begin{aligned}
f(-x) &= \frac{\sqrt[3]{-x}}{\sqrt[3]{1+x}+\sqrt[3]{1-x}} \\
&= \frac{-\sqrt[3]{x}}{\sqrt[3]{1-x}+\sqrt[3]{1+x}} \\
&= -f(x), \quad \forall\, x \in \mathbb{R}.
\end{aligned}$$

5.3. CALCULUS

Recall that the integral of any odd function on a symmetric interval around 0 is equal to 0. Then, we conclude that
$$\int_{-1}^{1} \frac{\sqrt[3]{x}}{\sqrt[3]{1-x} + \sqrt[3]{1+x}} \, dx = 0. \quad \square$$

Question 8. Compute the integral
$$\int (x^6 + x^3) \sqrt[3]{x^3 + 2} \, dx.$$

Answer: By factoring x out of $x^6 + x^3$ and moving it under the cube root, we obtain that

$$\int (x^6 + x^3) \sqrt[3]{x^3 + 2} \, dx$$
$$= \int (x^5 + x^2) \cdot x \sqrt[3]{x^3 + 2} \, dx$$
$$= \int (x^5 + x^2) \sqrt[3]{x^6 + 2x^3} \, dx. \quad (5.38)$$

We now use the substitution $u = x^6 + 2x^3$. Then, $du = 6(x^5 + x^2) \, dx$ and therefore $dx = \frac{du}{6(x^5 + x^2)}$. Thus,

$$\int (x^5 + x^2) \sqrt[3]{x^6 + 2x^3} \, dx$$
$$= \int (x^5 + x^2) \sqrt[3]{u} \, \frac{du}{6(x^5 + x^2)}$$
$$= \int \frac{1}{6} \sqrt[3]{u} \, du$$
$$= \frac{1}{6} \cdot \frac{u^{\frac{4}{3}}}{\frac{4}{3}} + C$$
$$= \frac{1}{8} (x^6 + 2x^3)^{\frac{4}{3}} + C, \quad (5.39)$$

where C is a real constant.

From (5.38) and (5.39), we conclude that

$$\int (x^6 + x^3) \sqrt[3]{x^3 + 2}\, dx = \frac{1}{8} \left(x^6 + 2x^3\right)^{\frac{4}{3}} + C. \quad \square$$

Question 9. Compute the integral

$$\int_0^{\frac{\pi}{2}} \ln(\sin x)\, dx.$$

Answer: Let

$$\mathcal{I} = \int_0^{\frac{\pi}{2}} \ln(\sin x)\, dx. \tag{5.40}$$

We will use the substitution $u = \frac{\pi}{2} - x$. Then, $x = 0$ corresponds to $u = \frac{\pi}{2}$, $x = \frac{\pi}{2}$ corresponds to $u = 0$, $x = \frac{\pi}{2} - u$, and therefore $dx = -du$. Thus,

$$\begin{aligned}
\mathcal{I} &= \int_0^{\frac{\pi}{2}} \ln(\sin(x))\, dx \\
&= \int_{\frac{\pi}{2}}^{0} \ln(\cos(u))\, (-du) \\
&= \int_0^{\frac{\pi}{2}} \ln(\cos(u))\, du \\
&= \int_0^{\frac{\pi}{2}} \ln(\cos(x))\, dx, \tag{5.41}
\end{aligned}$$

where for the last equality we changed the integration variable from u to x.

By adding (5.40) and (5.41), we obtain that

$$\begin{aligned}
2\mathcal{I} &= \int_0^{\frac{\pi}{2}} \ln(\sin(x)) + \ln(\cos(x))\, dx \\
&= \int_0^{\frac{\pi}{2}} \ln(\sin(x)\cos(x))\, dx. \tag{5.42}
\end{aligned}$$

5.3. CALCULUS

Recall that $\sin(2x) = 2\sin(x)\cos(x)$ and write (5.42) as

$$\begin{aligned}
2\mathcal{I} &= \int_0^{\frac{\pi}{2}} \ln\left(\frac{1}{2}\sin(2x)\right) dx \\
&= \int_0^{\frac{\pi}{2}} \ln\left(\frac{1}{2}\right) dx + \int_0^{\frac{\pi}{2}} \ln(\sin(2x)) dx \\
&= -\frac{\pi}{2}\ln 2 + \int_0^{\frac{\pi}{2}} \ln(\sin(2x)) dx, \qquad (5.43)
\end{aligned}$$

since $\ln\left(\frac{1}{2}\right) = -\ln 2$.

By using the substitution $w = 2x$, we find that

$$\begin{aligned}
&\int_0^{\frac{\pi}{2}} \ln(\sin(2x)) dx \\
&= \frac{1}{2}\int_0^{\pi} \ln(\sin(w)) dw \\
&= \frac{1}{2}\int_0^{\frac{\pi}{2}} \ln(\sin(w)) dw + \frac{1}{2}\int_{\frac{\pi}{2}}^{\pi} \ln(\sin(w)) dw \\
&= \frac{\mathcal{I}}{2} + \frac{1}{2}\int_{\frac{\pi}{2}}^{\pi} \ln(\sin(w)) dw \qquad (5.44)
\end{aligned}$$

Furthermore, by using the substitution $u = \pi - w$, we obtain that

$$\begin{aligned}
&\int_{\frac{\pi}{2}}^{\pi} \ln(\sin(w)) dw \\
&= \int_{\frac{\pi}{2}}^{0} \ln(\sin(\pi - u))(-du) \\
&= \int_0^{\frac{\pi}{2}} \ln(\sin(u)) du \\
&= \mathcal{I}, \qquad (5.45)
\end{aligned}$$

where we used the fact that $\sin(\pi - u) = \sin(u)$.

From (5.44) and (5.45), it follows that

$$\int_0^{\frac{\pi}{2}} \ln\left(\sin(2x)\right) dx = \frac{\mathcal{I}}{2} + \frac{\mathcal{I}}{2} = \mathcal{I}, \qquad (5.46)$$

and, from (5.43) and (5.46), we conclude that

$$2\mathcal{I} = -\frac{\pi}{2}\ln 2 + \mathcal{I}.$$

Thus, $\mathcal{I} = -\frac{\pi}{2}\ln 2$ and therefore

$$\int_0^{\frac{\pi}{2}} \ln\left(\sin x\right) dx = -\frac{\pi}{2}\ln 2. \quad \square$$

Question 10. Compute the integral

$$\int_0^{\frac{\pi}{4}} \ln\left(1 + \tan x\right) dx.$$

Answer: Let

$$\mathcal{I} = \int_0^{\frac{\pi}{4}} \ln\left(1 + \tan x\right) dx.$$

We use the substitution $y = \frac{\pi}{4} - x$. Then, $x = 0$ corresponds to $y = \frac{\pi}{4}$, $x = \frac{\pi}{4}$ corresponds to $y = 0$, $x = \frac{\pi}{4} - y$ and therefore $dx = -dy$. Thus,

$$\begin{aligned}
\mathcal{I} &= \int_{\frac{\pi}{4}}^0 \ln\left(1 + \tan\left(\frac{\pi}{4} - y\right)\right)(-dy) \\
&= \int_0^{\frac{\pi}{4}} \ln\left(1 + \tan\left(\frac{\pi}{4} - y\right)\right) dy.
\end{aligned}$$

Using the formula for the tangent of the sum and the fact that $\tan\frac{\pi}{4} = 1$, we obtain

$$\begin{aligned}
\tan\left(\frac{\pi}{4} - y\right) &= \frac{\tan\left(\frac{\pi}{4}\right) - \tan y}{1 + \tan\left(\frac{\pi}{4}\right)\tan y} \\
&= \frac{1 - \tan y}{1 + \tan y}.
\end{aligned}$$

5.3. CALCULUS

Therefore,

$$\begin{aligned}
\mathcal{I} &= \int_0^{\frac{\pi}{4}} \ln\left(1 + \frac{1-\tan y}{1+\tan y}\right) dy \\
&= \int_0^{\frac{\pi}{4}} \ln\left(\frac{2}{1+\tan y}\right) dy \\
&= \int_0^{\frac{\pi}{4}} (\ln 2 - \ln(1+\tan y))\, dy \\
&= \frac{\pi}{4}\ln 2 - \int_0^{\frac{\pi}{4}} \ln(1+\tan y)\, dy \\
&= \frac{\pi}{4}\ln 2 - \mathcal{I}. \qquad (5.47)
\end{aligned}$$

From (5.47), it follows that $\mathcal{I} = \frac{\pi}{8}\ln 2$ and therefore

$$\int_0^{\frac{\pi}{4}} \ln(1+\tan x)\, dx = \frac{\pi}{8}\ln 2. \quad \square$$

Question 11. Compute the integral

$$\int_0^1 x^x\, dx.$$

Answer: Using the Taylor series

$$e^z = \sum_{j=0}^{\infty} \frac{z^j}{j!}, \quad \forall\, z \in \mathbb{R},$$

for the exponential function, we obtain that

$$\begin{aligned}
\int_0^1 x^x\, dx &= \int_0^1 e^{x \ln x}\, dx \\
&= \int_0^1 \sum_{j=0}^{\infty} \frac{x^j (\ln x)^j}{j!}\, dx.
\end{aligned}$$

290 CHAPTER 5. MATHEMATICAL TIDBITS

We are allowed to change the order of the integral and the summation. However, we would need some measure theory to formally prove that such change is allowed. We will omit the details here. After the summation and the integral are switched, we obtain

$$\int_0^1 x^x\,dx = \sum_{j=0}^\infty \frac{1}{j!}\int_0^1 x^j (\ln x)^j\,dx$$
$$= \sum_{j=0}^\infty \frac{1}{j!} I(j,j), \qquad (5.48)$$

where $\mathcal{I}(m,n)$ is defined for nonnegative integers m and n as follows:

$$\mathcal{I}(m,n) = \int_0^1 x^m (\ln x)^n\,dx. \qquad (5.49)$$

Note that

$$\mathcal{I}(m,0) = \int_0^1 x^m\,dx = \frac{1}{m+1}. \qquad (5.50)$$

To calculate $\mathcal{I}(m,n)$, we use the integration by parts

$$\int u\,dv = uv - \int v\,du$$

with $u = (\ln x)^n$, $dv = x^m\,dx$, $du = n(\ln x)^{n-1}\cdot\frac{1}{x}\,dx$, and $v = \frac{x^{m+1}}{m+1}$. Then, we can write (5.49) as

$$\mathcal{I}(m,n)$$
$$= \int_0^1 x^m (\ln x)^n\,dx$$
$$=$$
$$= \frac{x^{m+1}}{m+1}\cdot(\ln x)^n\Big|_0^1 - \frac{n}{m+1}\int_0^1 x^m(\ln x)^{n-1}\,dx$$
$$= -\frac{n}{m+1}\mathcal{I}(m,n-1), \qquad (5.51)$$

5.3. CALCULUS

where the last equality follows from the fact that, for any integers $m, n \geq 0$,

$$\lim_{x \to 0} x^{m+1} (\ln x)^n = 0. \qquad (5.52)$$

We will prove this statement later on.

By iterating (5.51), we obtain that

$$\begin{aligned}
\mathcal{I}(m,n) &= -\frac{n}{m+1}\mathcal{I}(m, n-1) \\
&= -\frac{n}{m+1}\left(-\frac{n-1}{m+1}\mathcal{I}(m, n-2)\right) \\
&= \frac{n(n-1)}{(m+1)^2}\mathcal{I}(m, n-2) \\
&= \frac{n(n-1)}{(m+1)^2}\left(-\frac{n-2}{m+1}\mathcal{I}(m, n-3)\right) \\
&= -\frac{n(n-1)(n-2)}{(m+1)^3}\mathcal{I}(m, n-3) \\
&= \cdots \\
&= (-1)^n \frac{n!}{(m+1)^n}\mathcal{I}(m, 0) \\
&= (-1)^n \frac{n!}{(m+1)^{n+1}},
\end{aligned}$$

where the last equality follows from (5.50). Thus,

$$\mathcal{I}(j,j) = \frac{(-1)^j j!}{(j+1)^{j+1}}. \qquad (5.53)$$

From (5.48) and (5.53), we conclude that

$$\begin{aligned}
\int_0^1 x^x \, dx &= \sum_{j=0}^{\infty} \frac{1}{j!} \cdot \frac{(-1)^j j!}{(j+1)^{j+1}} \\
&= \sum_{j=0}^{\infty} \frac{(-1)^j}{(j+1)^{j+1}}. \qquad (5.54)
\end{aligned}$$

We now prove (5.52). Let $y = -\ln x$ for $x > 0$. Then, $x = e^{-y} = \frac{1}{e^y}$ and the limit from (5.52) becomes

$$\lim_{x \to 0} x^{m+1} (\ln x)^n = (-1)^n \lim_{y \to \infty} \frac{y^n}{e^{(m+1)y}} \qquad (5.55)$$

Let $z = (m+1)y$. Then, $y = \frac{z}{m+1}$ and

$$\lim_{y \to \infty} \frac{y^n}{e^{(m+1)y}} = \frac{1}{(m+1)^n} \lim_{z \to \infty} \frac{z^n}{e^z}. \qquad (5.56)$$

The last limit is equal to 0. This can be proved using L'Hopital's rule. We will prove it using sequeeze theorem. Since

$$e^z = \sum_{k=0}^{\infty} \frac{z^k}{k!} \geq \frac{z^{n+1}}{(n+1)!},$$

for all $z \geq 0$ and all positive integers n, it follows that

$$\frac{z^n}{e^z} \leq \frac{(n+1)!}{z}. \qquad (5.57)$$

From (5.56) and (5.57), and since y is positive, we conclude that

$$\begin{aligned}
0 &\leq \lim_{y \to \infty} \frac{y^n}{e^{(m+1)y}} \\
&\leq \frac{1}{(m+1)^n} \lim_{z \to \infty} \frac{(n+1)!}{z} \\
&= \frac{(n+1)!}{(m+1)^n} \lim_{z \to \infty} \frac{1}{z} \\
&= 0.
\end{aligned}$$

Thus,

$$\lim_{y \to \infty} \frac{y^n}{e^{(m+1)y}} = 0,$$

and, from (5.55), we obtain that

$$\lim_{x \to 0} x^{m+1} (\ln x)^n = 0$$

5.3. CALCULUS

and therefore (5.52) is established. □

Question 12. Compute the integral

$$\int_0^\infty e^{-xy} \cdot \frac{\sin(ax)}{x}\, dx.$$

Answer: Assume that $y > 0$ and denote the given integral we have to compute by $F(y, a)$:

$$F(y, a) = \int_0^\infty e^{-xy} \cdot \frac{\sin(ax)}{x}\, dx. \qquad (5.58)$$

Then,

$$\frac{\partial}{\partial y} F(y, a) = -\int_0^\infty e^{-xy} \cdot \sin(ax)\, dx. \qquad (5.59)$$

Let

$$I = \int_0^\infty e^{\beta x} \sin(\alpha x)\, dx,$$

where $\beta < 0$. By applying the integration by parts formula

$$\int A(x) B'(x)\, dx = A(x) B(x) - \int A'(x) B(x)\, dx \qquad (5.60)$$

for $A(x) = e^{\beta x}$ and $B(x) = \frac{-\cos(\alpha x)}{\alpha}$, it follows that

$$I = -\frac{e^{\beta x} \cos(\alpha x)}{\alpha}\bigg|_{x=0}^{x=\infty} + \frac{\beta}{\alpha} \int_0^\infty e^{\beta x} \cos(\alpha x)\, dx$$

$$\stackrel{\beta \leq 0}{=} \frac{1}{\alpha} + \frac{\beta}{\alpha} \int_0^\infty e^{\beta x} \cos(\alpha x)\, dx.$$

We use once again the integration by parts (5.60) with $A(x) = e^{\beta x}$ and $B(x) = \frac{\sin(\alpha x)}{\alpha}$ to find that

$$I = \frac{1}{\alpha} + \left.\frac{\beta e^{\beta x} \sin(\alpha x)}{\alpha^2}\right|_{x=0}^{x=\infty}$$
$$- \frac{\beta^2}{\alpha^2} \int_0^\infty e^{\beta x} \sin(\alpha x)\, dx$$
$$\stackrel{\beta \leq 0}{=} \frac{1}{\alpha} - \frac{\beta^2}{\alpha^2} I.$$

Thus,

$$I = \frac{1}{\alpha} - \frac{\beta^2}{\alpha^2} I, \tag{5.61}$$

and, by solving (5.61) for I, we obtain that

$$I = \int_0^\infty e^{\beta x} \sin(\alpha x)\, dx = \frac{\alpha}{\alpha^2 + \beta^2}. \tag{5.62}$$

The equation (5.59) becomes

$$\frac{\partial}{\partial y} F(y, a) = -\frac{a}{a^2 + y^2}. \tag{5.63}$$

This is an ordinary differential equation for the function $y \mapsto F(y, a)$ where the parameter a is fixed. Using that

$$\int \frac{a}{a^2 + y^2}\, dy = \arctan \frac{y}{a} + C,$$

for some constant C, we derive

$$F(y, a) = -\arctan \frac{y}{a} + C.$$

We will now determine the constant C. By placing $a = 0$ into (5.58) we conclude that $F(y, 0) = 0$. Since

$$\lim_{a \to 0_+} \arctan \frac{y}{a} = \frac{\pi}{2}$$

5.3. CALCULUS

and $F(y,0) = 0$ we conclude that $C = \frac{\pi}{2}$. Thus, the required integral is $\frac{\pi}{2} - \arctan \frac{y}{a}$.

If $y < 0$, the integral is divergent. To see this, we first use the change of variables $ax = t$ and write the integral as

$$\int_0^\infty e^{-xy} \cdot \frac{\sin(ax)}{x} \, dx = \int_0^\infty e^{-t\frac{y}{a}} \cdot \frac{\sin(t)}{t} \, dt$$
$$= \int_0^\infty e^{ct} \frac{\sin(t)}{t} \, dt,$$

where $c = -\frac{y}{a} > 0$ is a positive constant since $y < 0$ and $a > 0$.

In order to show that the integral $\int_0^\infty e^{ct} \frac{\sin(t)}{t} dt$ is divergent, we will prove that

$$\lim_{j \to \infty} \left| \int_{2j\pi}^{2(j+1)\pi} e^{ct} \frac{\sin(t)}{t} \, dt \right| = \infty,$$

where j is a positive integer. Note that

$$\int_{2j\pi}^{2(j+1)\pi} e^{ct} \frac{\sin(t)}{t} \, dt = \int_{2j\pi}^{(2j+1)\pi} e^{ct} \frac{\sin(t)}{t} \, dt \quad (5.64)$$
$$+ \int_{(2j+1)\pi}^{2(j+1)\pi} e^{ct} \frac{\sin(t)}{t} \, dt. (5.65)$$

By using the change of variables $t = 2\pi j + s_1$, we obtain that the integral from (5.64) can be written as

$$\int_{2j\pi}^{(2j+1)\pi} e^{ct} \frac{\sin(t)}{t} \, dt$$
$$= \int_0^\pi e^{c(2\pi j + s_1)} \frac{\sin(2\pi j + s_1)}{2\pi j + s_1} \, ds_1$$
$$= e^{2j\pi c} \int_0^\pi e^{cs_1} \frac{\sin(s_1)}{2\pi j + s_1} ds_1 \quad (5.66)$$
$$= e^{2j\pi c} \int_0^\pi e^{ct} \frac{\sin(t)}{2\pi j + t} \, dt, \quad (5.67)$$

where, for (5.66), we used the fact that $\sin(2\pi j + s_1) = \sin(s_1)$ since j is a positive integer, and, for (5.67), we used the notation t for the variable s_1.

Similarly, by using the change of variables $t = (2j+1)\pi + s_2$, the integral from (5.65) can be written as

$$\int_{(2j+1)\pi}^{2(j+1)\pi} e^{ct} \frac{\sin(t)}{t}\, dt$$

$$= \int_0^\pi e^{c((2j+1)\pi + s_2)} \frac{\sin((2j+1)\pi + s_2)}{(2j+1)\pi + s_2}\, ds_2$$

$$= e^{2j\pi c} \int_0^\pi e^{c\pi + cs_2} \frac{-\sin(s_2)}{(2j+1)\pi + s_2}\, ds_2 \qquad (5.68)$$

$$= -e^{2j\pi c} \int_0^\pi e^{c\pi + ct} \frac{\sin(t)}{(2j+1)\pi + t}\, dt, \qquad (5.69)$$

where, for (5.68), we used the fact that $\sin((2j+1)\pi + s_2) = -\sin(s_2)$ since j is a positive integer, and, for (5.69), we used the notation t for the variable s_2.

From (5.64–5.65) and using (5.67) and (5.69), we find that

$$\int_{2j\pi}^{2(j+1)\pi} e^{ct} \frac{\sin(t)}{t}\, dt \qquad (5.70)$$

$$= e^{2j\pi c} \int_0^\pi e^{ct} \frac{\sin(t)}{2\pi j + t}\, dt$$

$$\quad - e^{2j\pi c} \int_0^\pi e^{c\pi + ct} \frac{\sin(t)}{(2j+1)\pi + t}\, dt$$

$$= -e^{2j\pi c} \int_0^\pi e^{ct} \sin(t) \cdot D(t, j)\, dt,$$

where $D(t, j)$ is defined as

$$D(t, j) = \frac{e^{c\pi}}{(2j+1)\pi + t} - \frac{1}{2\pi j + t}. \qquad (5.71)$$

We will find a lower bound for $D(t, j)$. First, we will transform $D(t, j)$ by finding a common denominator of

5.3. CALCULUS

the two fractions.

$$D(t,j) = \frac{(e^{c\pi} - 1)(2\pi j + t) - \pi}{((2j+1)\pi + t) \cdot (2\pi j + t)}.$$

We will now use that $t > 0$ and obtain a lower bound for $D(t,j)$ that does not contain the variable t in the numerator.

$$D(t,j) > \frac{(e^{c\pi} - 1) \cdot 2\pi j - \pi}{((2j+1)\pi + t) \cdot (2\pi j + t)}.$$

Next, we will use the inequality $t \leq \pi$ and obtain a lower bound for $D(t,j)$ that does not depend on t.

$$\begin{aligned} D(t,j) &> \frac{(e^{c\pi} - 1) \cdot 2\pi j - \pi}{(2j+2)\pi \cdot (2j+1)\pi} \\ &= \frac{(e^{c\pi} - 1) \cdot 2j - 1}{2(j+1) \cdot (2j+1)\pi}. \end{aligned} \quad (5.72)$$

Let us denote by $\tilde{D}(j)$ the right-hand side of (5.72). More precisely,

$$\tilde{D}(j) = \frac{(e^{c\pi} - 1) \cdot 2j - 1}{2(j+1) \cdot (2j+1)\pi}. \quad (5.73)$$

We will now prove that there is a positive constant $\tilde{C}$ and a positive integer j_0 such that

$$j > j_0 \implies \tilde{D}(j) > \frac{\tilde{C}}{j}. \quad (5.74)$$

The number $\tilde{D}(j)$ satisfies

$$\begin{aligned} \tilde{D}(j) &= \frac{j \cdot \left[2(e^{c\pi} - 1) - \frac{1}{j}\right]}{j^2 \left[2\left(1 + \frac{1}{j}\right) \cdot \left(2 + \frac{1}{j}\right)\pi\right]} \\ &= \frac{1}{j} \cdot \frac{2(e^{c\pi} - 1) - \frac{1}{j}}{\left(1 + \frac{1}{j}\right) \cdot \left(2 + \frac{1}{j}\right) \cdot 2\pi}. \end{aligned} \quad (5.75)$$

We will use $1 + \frac{1}{j} < 2$ and $2 + \frac{1}{j} < 3$ to bound the denominator in (5.75). We obtain that for every $j \geq 1$ we have

$$\tilde{D}(j) \geq \frac{1}{j} \cdot \frac{2(e^{c\pi} - 1) - \frac{1}{j}}{12\pi}. \tag{5.76}$$

Let j_0 be the smallest positive integer bigger than $\frac{1}{e^{c\pi}-1}$, i.e. $j_0 = \lceil \frac{1}{e^{c\pi}-1} \rceil$. Then, for $j > j_0$, the numerator in (5.76) is bigger than $e^{c\pi} - 1$. Hence, for $j > j_0$ we have

$$\tilde{D}(j) > \frac{1}{j} \cdot \frac{e^{c\pi} - 1}{12\pi}. \tag{5.77}$$

Thus, if we define

$$j_0 = \left\lceil \frac{1}{e^{c\pi} - 1} \right\rceil \quad \text{and} \quad \tilde{C} = \frac{e^{c\pi} - 1}{12\pi},$$

the inequality $D(j) > \frac{\tilde{C}}{j}$ holds for every $j > j_0$. The absolute value of the integral from (5.70) can be bounded for every $j > j_0$ in the following way.

$$\left| \int_{2j\pi}^{2(j+1)\pi} e^{ct} \frac{\sin(t)}{t} \, dt \right| \tag{5.78}$$
$$> e^{2j\pi c} \cdot \frac{\tilde{C}}{j} \cdot \int_0^{\pi} e^{ct} \sin(t) \, dt.$$

The value of the last integral is a positive real number. Let us denote it by $\tilde{E}$, i.e.

$$\tilde{E} = \int_0^{\pi} e^{ct} \sin(t) \, dt.$$

Then, from (5.78), we obtain that for every $j > j_0$, the following inequality holds

$$\left| \int_{2j\pi}^{2(j+1)\pi} e^{ct} \frac{\sin(t)}{t} \, dt \right| > \tilde{C}\tilde{E} \, \frac{e^{2j\pi c}}{j}. \tag{5.79}$$

5.3. CALCULUS

where $\tilde{C}$ and $\tilde{E}$ are two positive real numbers.
Then, from (5.79), we conclude that

$$\lim_{j \to \infty} \left| \int_{2j\pi}^{2(j+1)\pi} e^{ct} \frac{\sin(t)}{t} \, dt \right|$$
$$\geq \lim_{j \to \infty} \tilde{C}\tilde{E} \frac{e^{2j\pi c}}{j} \geq \lim_{j \to \infty} \tilde{C}\tilde{E} \frac{(2j\pi c)^2}{2j} = \infty,$$

where we used the fact that $e^x > \frac{x^2}{2}$ for $x = 2j\pi c$.
This concludes the proof of the fact that the given integral is divergent if $y < 0$. $\square$

Question 13. Compute the integral

$$\int_0^1 \int_0^1 \frac{1}{1 - xy} \, dx \, dy.$$

Answer: Since $0 \leq x, y \leq 1$, we can expand $\frac{1}{1-xy}$ as the geometric series $\sum_{n=0}^{\infty} (xy)^n$ and obtain that

$$\begin{aligned}
\int_0^1 \int_0^1 \frac{1}{1 - xy} \, dx \, dy &= \int_0^1 \int_0^1 \sum_{n=0}^{\infty} (xy)^n \, dx \, dy \\
&= \sum_{n=0}^{\infty} \left(\int_0^1 x^n \, dx \right) \left(\int_0^1 y^n \, dy \right) \\
&= \sum_{n=0}^{\infty} \frac{1}{(n+1)^2} \\
&= \frac{\pi^2}{6},
\end{aligned}$$

where the last step follows from the fact that the Riemann zeta function corresponding to exponent 2 is $\frac{\pi^2}{6}$:

$$\zeta(2) = \sum_{k=1}^{\infty} \frac{1}{k^2} = \frac{\pi^2}{6}. \quad \square$$

CHAPTER 5. MATHEMATICAL TIDBITS

Question 14. Let $a > 0$, $b > 0$. Compute the integral

$$\int_0^\infty \frac{e^{-ax} - e^{-bx}}{x} \, dx.$$

Answer: Note that

$$\frac{e^{-ax} - e^{-bx}}{x} = \int_a^b e^{-xt} \, dt$$

and therefore

$$\int_0^\infty \frac{e^{-ax} - e^{-bx}}{x} \, dx = \int_0^\infty \int_a^b e^{-xt} \, dt \, dx. \quad (5.80)$$

Using Fubini's theorem to switch the order of integration in (5.80), we obtain

$$\begin{aligned}
\int_0^\infty \frac{e^{-ax} - e^{-bx}}{x} \, dx &= \int_a^b \int_0^\infty e^{-xt} \, dx \, dt \\
&= \int_a^b \left(-\frac{e^{-xt}}{t}\right)\bigg|_{x=0}^{x=\infty} dt \\
&= \int_a^b \frac{1}{t} \, dt = (\ln t)\bigg|_{t=a}^{t=b} \\
&= \ln\left(\frac{b}{a}\right). \quad \square
\end{aligned}$$

Question 15. Find the sum of the infinite series

$$\sum_{n=1}^\infty \frac{1}{2n^2 - n}.$$

Answer: First Solution: Notice that

$$\begin{aligned}
\sum_{n=1}^p \frac{1}{2n^2 - n} &= \sum_{n=1}^p \left(\frac{2}{2n-1} - \frac{1}{n}\right) \\
&= \sum_{n=1}^p \frac{2}{2n-1} - \sum_{n=1}^p \frac{1}{n}.
\end{aligned}$$

5.3. CALCULUS

We will add $\sum_{n=1}^{p} \frac{2}{2n}$ to each of the last two sums and obtain

$$\sum_{n=1}^{p} \frac{1}{2n^2 - n}$$
$$= \left(\sum_{n=1}^{p} \frac{2}{2n-1} + \sum_{n=1}^{p} \frac{2}{2n} \right) - \left(\sum_{n=1}^{p} \frac{1}{n} + \sum_{n=1}^{p} \frac{2}{2n} \right)$$
$$= 2 \sum_{n=1}^{2p} \frac{1}{n} - 2 \sum_{n=1}^{p} \frac{1}{n}. \tag{5.81}$$

Recall that

$$\lim_{m \to \infty} \sum_{n=1}^{m} \frac{1}{n} - \ln(m) = \gamma, \tag{5.82}$$

where $\gamma \approx 0.5772$ is Euler's constant.

We can write (5.81) as follows:

$$\sum_{n=1}^{p} \frac{1}{2n^2 - n}$$
$$= 2 \left(\sum_{n=1}^{2p} \frac{1}{n} - \ln(2p) \right) + 2\ln(2p)$$
$$\quad - 2 \left(\sum_{n=1}^{p} \frac{1}{n} - \ln(p) \right) - 2\ln(p)$$
$$= 2\ln(2) + 2 \left(\sum_{n=1}^{2p} \frac{1}{n} - \ln(2p) \right)$$
$$\quad - 2 \left(\sum_{n=1}^{p} \frac{1}{n} - \ln(p) \right). \tag{5.83}$$

By letting p go to infinity in (5.83) and using (5.82), we conclude that

$$\sum_{n=1}^{\infty} \frac{1}{2n^2 - n} = \lim_{p \to \infty} \sum_{n=1}^{p} \frac{1}{2n^2 - n}$$
$$= 2\ln(2) + 2\gamma - 2\gamma$$
$$= 2\ln(2).$$

Second Solution: Consider the function

$$f(x) = \sum_{n=1}^{\infty} \frac{x^n}{2n^2 - n}. \tag{5.84}$$

The derivative of the function f satisfies

$$f'(x) = \sum_{n=1}^{\infty} \frac{x^{n-1}}{2n - 1} = \frac{1}{\sqrt{x}} \sum_{n=1}^{\infty} \frac{\sqrt{x}^{2n-1}}{2n - 1}. \tag{5.85}$$

Denote by $g(x)$ the function given by

$$g(x) = \sum_{n=1}^{\infty} \frac{x^{2n-1}}{2n - 1}. \tag{5.86}$$

The derivative of g satisfies

$$g'(x) = \sum_{n=1}^{\infty} x^{2n-2}.$$

The last sum is a geometric series. Therefore,

$$g'(x) = \frac{1}{1 - x^2}.$$

Hence, the function g is an anti-derivative of $\frac{1}{1-x^2}$. Using that $\frac{1}{1-x^2} = \frac{1}{2} \cdot \frac{1}{1+x} + \frac{1}{2} \cdot \frac{1}{1-x}$, we obtain that g must satisfy

$$g(x) = \int \frac{1}{1 - x^2}\, dx = \frac{\ln|1 + x|}{2} - \frac{\ln|1 - x|}{2} + C,$$

5.3. CALCULUS

for some real constant C. From (5.86) we see that $g(0) = 0$. Therefore, $C = 0$. The equation (5.85) can now be written as

$$\begin{aligned} f'(x) &= \frac{1}{\sqrt{x}} g\left(\sqrt{x}\right) \\ &= \frac{\ln|1+\sqrt{x}|}{2\sqrt{x}} - \frac{\ln|1-\sqrt{x}|}{2\sqrt{x}}. \end{aligned} \quad (5.87)$$

The antiderivatives of the last two functions can be found using the substitution $\sqrt{x} = u$.

$$\begin{aligned} & \int \frac{\ln|1+\sqrt{x}|}{2\sqrt{x}} \, dx \\ =\ & \int \ln|1+u| \, du \\ =\ & (1+u)\ln(1+u) - u + D \\ =\ & \left(1+\sqrt{x}\right)\ln\left(1+\sqrt{x}\right) - \sqrt{x} + D, \end{aligned}$$

for some real constant D. In a similar way we obtain

$$\begin{aligned} & \int \frac{\ln|1-\sqrt{x}|}{2\sqrt{x}} \, dx \\ =\ & -\left(1-\sqrt{x}\right)\ln\left(1-\sqrt{x}\right) - \sqrt{x} + E, \end{aligned}$$

for some real constant E. From (5.87) we now obtain

$$\begin{aligned} & f(x) \qquad\qquad\qquad\qquad\qquad\qquad (5.88) \\ =\ & \left(1+\sqrt{x}\right)\ln\left(1+\sqrt{x}\right) + \left(1-\sqrt{x}\right)\ln\left(1-\sqrt{x}\right) + G, \end{aligned}$$

where G is a real constant (the constant G is equal to $D - E$). From (5.84) we see that $f(0) = 0$. This implies that $G = 0$. The function f is defined for every $x < 1$. We can evaluate the limit as $x \to 1$. We use the substitution $y = -\ln\left(1-\sqrt{x}\right)$ to evaluate the limit

$$\lim_{x \to 1} \left(1-\sqrt{x}\right)\ln\left(1-\sqrt{x}\right) = \lim_{y \to \infty} y e^{-y} = 0.$$

The equation (5.88) now implies that

$$\lim_{x \to 1} f(x) = 2\ln 2.$$

Thus, the sum of the given infinite series is $2\ln 2$. □

Question 16. Let k be a positive integer. Calculate

$$\sum_{n_1=1}^{\infty} \cdots \sum_{n_k=1}^{\infty} \frac{1}{n_1 \cdot \ldots \cdot n_k \cdot (n_1 + \ldots + n_k + 1)}.$$

Answer: Denote by S the summation above. Then,

$$\begin{aligned} S &= \sum_{n_1=1}^{\infty} \sum_{n_2=1}^{\infty} \cdots \sum_{n_k=1}^{\infty} \frac{1}{n_1 \cdot n_2 \cdot \ldots \cdot n_k} \times \\ &\quad \times \frac{1}{n_1 + n_2 + \ldots + n_k + 1} \\ &= \sum_{n_1=1}^{\infty} \sum_{n_2=1}^{\infty} \cdots \sum_{n_k=1}^{\infty} \frac{1}{n_1 \cdot n_2 \cdot \ldots \cdot n_k} \times \\ &\quad \times \int_0^1 x^{n_1+n_2+\ldots+n_k} \, dx \\ &= \int_0^1 \sum_{n_1=1}^{\infty} \frac{x^{n_1}}{n_1} \sum_{n_2=1}^{\infty} \frac{x^{n_2}}{n_2} \cdots \sum_{n_k=1}^{\infty} \frac{x^{n_k}}{n_k} \, dx. \quad (5.89) \end{aligned}$$

Recall that the Taylor series expansion of the function $\ln(1-x)$ is

$$\ln(1-x) = -\sum_{n=1}^{\infty} \frac{x^n}{n}.$$

Then, (5.89) becomes

$$S = \int_0^1 (-\ln(1-x))^k \, dx. \quad (5.90)$$

5.3. CALCULUS

Using the substitution $u = -\ln(1-x) \iff e^{-u} = 1-x$, the last integral in (5.90) becomes

$$\begin{aligned} S &= \int_0^1 (-\ln(1-x))^k \, dx = \int_0^\infty u^k e^{-u} \, du \\ &= k!, \end{aligned}$$

where the last equality comes from the following property of the Gamma function

$$\Gamma(k+1) = \int_0^\infty u^k e^{-u} \, du = k!. \quad \square$$

Question 17. Find a function that is equal to the sum of all of its derivatives.

Answer: Denote by $f(x)$ an infinitely differentiable function such that

$$f(x) = \sum_{k=1}^\infty f^{(k)}(x), \quad \forall \, x \in \mathbb{R}. \quad (5.91)$$

By differentiating (5.91), we obtain that

$$\begin{aligned} f'(x) &= \sum_{k=1}^\infty f^{(k+1)}(x) \\ &= \sum_{k=2}^\infty f^{(k)}(x) \\ &= \sum_{k=1}^\infty f^{(k)}(x) - f'(x) \\ &= f(x) - f'(x), \quad \forall \, x \in \mathbb{R}, \quad (5.92) \end{aligned}$$

where, for the last equality, we used (5.91).

From (5.92), it follows that the function $f(x)$ satisfies

$$f'(x) = \frac{1}{2} f(x), \quad \forall \, x \in \mathbb{R}. \quad (5.93)$$

The solution of the the first order linear differential equation (5.93) is

$$f(x) = Ce^{\frac{x}{2}},$$

where C is a fixed real constant. □

Question 18. Find the nonnegative solutions to the differential equation

$$yy'' + (y')^2 = yy'e^x.$$

Answer: Any constant nonnegative function is a solution to the given differential equation. The non-constant solutions to

$$yy'' + (y')^2 = yy'e^x. \quad (5.94)$$

are non–elementary functions related to the function[1]

$$\text{Ei}(x) = \int_{-\infty}^{x} \frac{e^t}{t} dt. \quad (5.95)$$

Let us introduce the substitution $z = yy'$. Then,

$$z' = yy'' + (y')^2,$$

and the differential equation (5.94) can be written as

$$z' = ze^x. \quad (5.96)$$

Then, it follows from (5.96) that

$$\left(ze^{-e^x}\right)' = (z' - ze^x)e^{-e^x} = 0,$$

---
[1] The function $\text{Ei}(x)$ given by (5.95) is the exponential integral function, one of the most frequently occurring non-elementary functions.

5.3. CALCULUS

and therefore we conclude that ze^{-e^x} is a constant. Denote this constant by $\frac{D}{2}$. Then, $ze^{-e^x} = \frac{D}{2}$ which can be written as

$$z = \frac{D}{2} e^{e^x}. \tag{5.97}$$

Recall that $z = yy'$ and note that

$$\frac{1}{2}\left(y^2\right)' = yy' = z. \tag{5.98}$$

From (5.97) and (5.98), we obtain that

$$\left(y^2\right)' = De^{e^x}. \tag{5.99}$$

Let

$$g(x) = \int_0^x e^{e^t}\, dt. \tag{5.100}$$

The equation (5.99) can now be written as

$$\left(y^2\right)' = Dg'(x),$$

which is equivalent to

$$\left(y^2 - Dg(x)\right)' = 0.$$

Then,

$$y^2(x) = Dg(x) + E, \tag{5.101}$$

for some constant E.

We will conclude by expressing $g(x)$ in terms of the function $\text{Ei}(x)$ from (5.95). The substitution $u = e^t$ transforms the integral from the right-hand side of (5.100) to

$$g(x) = \int_1^{e^x} \frac{e^u}{u}\, du = \text{Ei}\left(e^x\right) - \text{Ei}(1). \tag{5.102}$$

Therefore, any nonnegative solution y satisfies

$$y(x) = \sqrt{D\text{Ei}\left(e^x\right) + E - D\text{Ei}(1)}, \tag{5.103}$$

where D and E are real constants. The number $E - D\text{Ei}(1)$ is also a constant. If we denote this constant by G, then the function y can be written as

$$y(x) = \sqrt{D\text{Ei}(e^x) + G}. \qquad (5.104)$$

In the last equation D and G are constants and Ei is the exponential integral given by (5.95). $\square$

Question 19. When the clock strikes midnight, n bugs located at the vertices of a regular n-gon of unit side length begin crawling at equal speeds, in the clockwise directions, and directly towards the adjacent bug. They continue to do so until they all finally meet at the center of the polygon.

What is the total distance traveled by each bug? How many times does each bug spiral around the center?

Answer: To begin with, we change the problem into one in which bugs rotate in counterclockwise direction. We will solve the problem using polar coordinates, and we really do not want to re-invent the trigonometry in which sin and cos have their roles interchanged.

Denote by O the center of the n-gon and by denote by $\vec{r_i}(t) = \langle x_i(t), y_i(t) \rangle$ the position of the bug i at time t, for $i = 1 : n$. Due to symmetry, at any time, the bugs form a regular n-gon and are equidistant from the center O. Let $r(t)$ be the distance between each of the bugs from the center O. Denote by $\alpha_i(t)$ the angle between $\vec{r_i}(t) = \langle x_i(t), y_i(t) \rangle$ and the x-axis. Then,

$$\vec{r_i}(t) = \langle r(t)\cos\alpha_i(t), r(t)\sin\alpha_i(t) \rangle. \qquad (5.105)$$

We will choose the x–axis in such a way that $\alpha_1(0) = 0$. We will also make the convention that $\alpha_i(0) \in [0, 2\pi)$, for $i = 1 : n$; nonetheless, as t starts increasing, the angles

5.3. CALCULUS

$\alpha_i(t)$ will exit the canonical interval $[0, 2\pi)$. Due to the bugs moving in counter-clockwise direction, we have that

$$\alpha_{i+1}(t) - \alpha_i(t) = \frac{2\pi}{n},$$

for every i and every t. We will now determine the functions $r(t)$ and $\alpha_i(t)$ by establishing equations that correspond to the requirements that the bugs crawl directly towards the adjacent bug and that the bugs move at the same speed, respectively:

(1) the velocity vector $\vec{r_i'}(t)$ of the bug i is parallel to $\vec{r_{i+1}}(t) - \vec{r_i}(t)$;

(2) the speed $\left\|\vec{r_i'}(t)\right\|$ is equal to 1.

The velocity vector $\vec{r_i'}(t)$ is given by

$$\vec{r_i'}(t) = \langle x_i'(t), y_i'(t) \rangle,$$

where

$$x_i'(t) = r'(t)\cos\alpha_i(t) - r(t)\sin\alpha_i(t) \cdot \alpha_i'(t); \quad (5.106)$$
$$y_i'(t) = r'(t)\sin\alpha_i(t) + r(t)\cos\alpha_i(t) \cdot \alpha_i'(t). \quad (5.107)$$

The requirement (2) corresponds to

$$\left(x_i'(t)\right)^2 + \left(y_i'(t)\right)^2 = 1. \quad (5.108)$$

Using the formulas (5.106) and (5.107) for $x_i'(t)$ and $y_i'(t)$, respectively, the equation (5.108) becomes

$$\left(r'(t)\right)^2 + \left(r(t)\right)^2 \cdot \left(\alpha_i'(t)\right)^2 = 1. \quad (5.109)$$

From (5.105), we obtain that

$$\vec{r_{i+1}}(t) - \vec{r_i}(t) = \langle x_{i+1}(t) - x_i(t), y_{i+1}(t) - y_i(t) \rangle$$

CHAPTER 5. MATHEMATICAL TIDBITS

and

$$x_{i+1}(t) - x_i(t)$$
$$= r(t) \cdot (\cos \alpha_{i+1}(t) - \cos \alpha_i(t))$$
$$= -r(t) \cdot 2 \sin\left(\frac{\alpha_{i+1}(t) + \alpha_i(t)}{2}\right) \times$$
$$\times \sin\left(\frac{\alpha_{i+1}(t) - \alpha_i(t)}{2}\right)$$
$$= -2r(t) \sin\left(\alpha_i(t) + \frac{\pi}{n}\right) \cdot \sin\frac{\pi}{n};$$

$$y_{i+1}(t) - y_i(t)$$
$$= r(t) \cdot (\sin \alpha_{i+1}(t) - \sin \alpha_i(t))$$
$$= r(t) \cdot 2 \cos\left(\frac{\alpha_{i+1}(t) + \alpha_i(t)}{2}\right) \times$$
$$\times \sin\left(\frac{\alpha_{i+1}(t) - \alpha_i(t)}{2}\right)$$
$$= 2r(t) \cos\left(\alpha_i(t) + \frac{\pi}{n}\right) \cdot \sin\frac{\pi}{n}.$$

The requirement that $\overrightarrow{r'_i}(t)$ has to be parallel to $\overrightarrow{r_{i+1}}(t) - \overrightarrow{r_i}(t)$ can be written as

$$\frac{r'(t) \cos \alpha_i(t) - r(t) \sin \alpha_i(t) \cdot \alpha'_i(t)}{-2r(t) \sin\left(\alpha_i(t) + \frac{\pi}{n}\right) \cdot \sin\frac{\pi}{n}}$$
$$= \frac{r'(t) \sin \alpha_i(t) + r(t) \cos \alpha_i(t) \cdot \alpha'_i(t)}{2r(t) \cos\left(\alpha_i(t) + \frac{\pi}{n}\right) \cdot \sin\frac{\pi}{n}}$$

After we cross–multiply, cancel $2r(t) \sin \frac{\pi}{n}$, and separate the terms that contain $r'(t)$ from the terms that contain $r(t)\alpha'_i(t)$, we obtain that

$$r'(t) \cdot \cos\left[\left(\alpha_i(t) + \frac{\pi}{n}\right) - \alpha_i(t)\right]$$
$$= r(t)\alpha'_i(t) \sin\left[\alpha_i(t) - \left(\alpha_i(t) + \frac{\pi}{n}\right)\right].$$

5.3. CALCULUS

In other words,

$$r'(t)\cos\frac{\pi}{n} = -r(t)\alpha'_i(t)\sin\frac{\pi}{n},$$

which implies that

$$r(t)\alpha'_i(t) = -r'(t)\cot\frac{\pi}{n}. \qquad (5.110)$$

From (5.109) and (5.110), we find that

$$\left(r'(t)\right)^2\left(1+\cot^2\frac{\pi}{n}\right) = 1.$$

The last equation is equivalent to

$$\left(r'(t)\right)^2\left(1+\frac{\cos^2\frac{\pi}{n}}{\sin^2\frac{\pi}{n}}\right) = 1$$
$$\iff \left(r'(t)\right)^2 = \sin^2\frac{\pi}{n}. \qquad (5.111)$$

Since the distance from the origin is decreasing, $r'(t) < 0$ and we conclude from (5.111) that

$$r'(t) = -\sin\frac{\pi}{n}. \qquad (5.112)$$

Thus, the distance from the origin satisfies

$$r(t) = R - t\cdot\sin\frac{\pi}{n}, \qquad (5.113)$$

where $R = r(0)$ is the radius of the circle circumscribed around the polygon.

The equation (5.113) is sufficient to evaluate the total time t that the bugs spend traveling. The bugs are traveling until they achieve $r(t) = 0$. This happens for

$$t = \frac{R}{\sin\frac{\pi}{n}}.$$

Since each bug travels at the speed 1, the distance a bug travels is
$$d = t \cdot 1 = \frac{R}{\sin \frac{\pi}{n}}. \tag{5.114}$$
The polygon has the side-length 1, hence the radius of its circumcircle is
$$R = \frac{1}{2 \sin \frac{\pi}{n}}. \tag{5.115}$$
The distance that each bug travels is
$$d = \frac{1}{2 \sin^2 \frac{\pi}{n}}.$$
We now determine how many times each bug goes around the circle. The required number of times n satisfies
$$n = \frac{\alpha_i(t) - \alpha_i(0)}{2\pi}.$$
From (5.110) and (5.112), it follows that
$$\begin{aligned}\alpha_i'(t) &= \frac{-r'(t) \cot \frac{\pi}{n}}{r(t)} = \frac{\sin \frac{\pi}{n} \cdot \cot \frac{\pi}{n}}{R - t \sin \frac{\pi}{n}} \\ &= \frac{\cos \frac{\pi}{n}}{R - t \sin \frac{\pi}{n}}.\end{aligned} \tag{5.116}$$
Using the Fundamental Theorem of Calculus, we find from (5.116) that
$$\begin{aligned}\alpha_i(t) &= \alpha_i(0) + \int_0^t \frac{\cos \frac{\pi}{n}}{R - u \sin \frac{\pi}{n}} \, du \\ &= \alpha_i(0) - \frac{\cos \frac{\pi}{n}}{\sin \frac{\pi}{n}} \cdot \ln \left| R - u \sin \frac{\pi}{n} \right| \Big|_{u=0}^{u=t} \\ &= \alpha_i(0) + \frac{\cos \frac{\pi}{n}}{\sin \frac{\pi}{n}} \cdot \ln \frac{R}{R - t \sin \frac{\pi}{n}}. \end{aligned} \tag{5.117}$$
By letting $t \to \frac{R}{\sin \frac{\pi}{n}}$ in (5.117), we obtain that $\alpha_i(t) \to \infty$. Therefore, we conclude that each bug will make infinitely many rotations around the center. $\square$

5.4 Linear Algebra

Question 1. Let A and B be two square matrices of the same order that satisfy

$$A^{-1} + B^{-1} = (A+B)^{-1}.$$

Show that $\det(A) = \det(B)$.

Answer: By multiplying

$$A^{-1} + B^{-1} = (A+B)^{-1}$$

by $A+B$ to the left, we obtain that

$$\begin{aligned} I &= (A+B)\left(A^{-1} + B^{-1}\right) \\ &= I + AB^{-1} + BA^{-1} + I \end{aligned}$$

and therefore

$$I + AB^{-1} + BA^{-1} = 0. \tag{5.118}$$

Let $M = AB^{-1}$. Then,

$$M^{-1} = \left(AB^{-1}\right)^{-1} = \left(B^{-1}\right)^{-1} A^{-1} = BA^{-1},$$

and (5.118) becomes

$$I + M + M^{-1} = 0. \tag{5.119}$$

Multiply (5.119) by M to obtain that

$$M^2 + M + I = 0, \tag{5.120}$$

and multiply (5.120) by $M - I$ to obtain that

$$M^3 - I = 0.$$

Thus, $M^3 = I$. By taking the determinants we obtain that $\det(M)^3 = 1$ and therefore

$$\det(M) = 1. \tag{5.121}$$

Recall that $M = AB^{-1}$. Then, $MB = A$ and using (5.121) we conclude that

$$\det(A) = \det(M)\det(B) = \det(B). \quad \square$$

Question 2. Let M be a triangular $n \times n$ matrix with all the entries on or above the main diagonal equal to 1. Find a quick way to compute M^k, for $k \geq 1$.

Answer: Consider the directed graph G with n vertices labeled $V_1, V_2, \ldots, V_n$. The edge goes from vertex V_i to V_j if and only if $i \leq j$. The $n \times n$ incidence matrix of the graph G has the entry (i,j) equal to 1 if and only if there is an edge between V_i and V_j; otherwise, the entry (i,j) is equal to 0. Note that our given matrix M is exactly equal to the incidence matrix of the graph G.

We will use the following result:

If M is the incidence matrix of a graph, then $M^k(i,j)$ (i.e., the (i,j) entry of the matrix M^k) is equal to the total number of paths of length k from the vertex i to the vertex j, for $1 \leq i, j \leq n$.

The proof of this result falls into the category of math proofs with the following three properties: 1) easy to derive; 2) easy to write down; 3) difficult to read if someone else wrote it. So, instead of proving this result, we refer the reader to, for example, Proposition 8.6.7 on page 455 from [8].

Thus, $M^k(i,j)$ is equal to the number of paths of length k from the vertex V_i to the vertex V_j. Each such path is of the form

$$(X_0, X_1, \ldots, X_k), \text{ where}$$
$$X_0, X_1, \ldots, X_k \in \{V_1, V_2, \ldots, V_n\},$$
$$X_0 = V_i, \quad X_k = V_j,$$

5.4. LINEAR ALGEBRA

and the indices of X_0, X_1, ..., X_k form a non-decreasing sequence.

Since the vertices V_i and V_j are connected if and only if $i \leq j$, the number $M^k(i,j)$ is equal to the number of non-decreasing sequences $(x_0, \ldots, x_k)$ whose terms are from $\{1, \ldots, n\}$ such that $x_0 = i$ and $x_k = j$. Let $y_1 = x_1 - x_0$, $y_2 = x_2 - x_1$, ..., $y_k = x_k - x_{k-1}$. The numbers y_1, ..., y_k are non-negative and their sum is equal to $j - i$. There is a one-to-one correspondence between $(x_0, \ldots, x_k)$ and $(y_0, \ldots, y_k)$. Counting how many k-tuples $(y_0, \ldots, y_k)$ there are is equivalent to counting the number of ways of distributing $j - i$ apples to k people.

We will prove the following result:

Proposition. Let α and k be two integers that satisfy $\alpha \geq 0$ and $k > 0$. Let $A(\alpha, k)$ be the number of ways to distribute α identical apples to k people. The number $A(\alpha, k)$ satisfies

$$A(\alpha, k) = \binom{\alpha + k - 1}{k - 1}. \qquad (5.122)$$

Proof. Denote by S the set of all possible ways to distribute α apples to k people. We can represent S as the collection of all ordered sequences of length k whose components add up to α. For example, if we are giving 15 apples to 4 people, then the sequence $(7, 0, 4, 4)$ would correspond to the situation in which the first person gets 7 apples, the second gets 0, and the third and fourth get 4 apples each. We may now write:

$$S = \{(y_1, \ldots, y_k) : y_1, \ldots, y_k \in \mathbb{N}_0, y_1 + \cdots + y_k = \alpha\};$$

here, $\mathbb{N}_0$ is the set of all the nonnegative integers and includes 0. Consider the set T of all sequences of length $\alpha + k - 1$ that consist of α zeroes and $k - 1$ ones. For each

$(y_1, \ldots, y_k) \in S$ we define $f(y_1, \ldots, y_k)$ as follows:

$$f(y_1, \ldots, y_k) = \underbrace{00\ldots0}_{y_1}1\underbrace{00\ldots0}_{y_2}1\ldots1\underbrace{00\ldots0}_{y_k}.$$

Then, $f : S \to T$ is a bijection and therefore $|S| = |T|$. Moreover, finding the number of elements of T is the same problem as starting with $\alpha+k-1$ zeroes and then choosing $k-1$ of those zeroes and turning them into ones. This can be done in $\binom{\alpha+k-1}{k-1}$ ways. This completes the proof of the proposition. □

Since $M^k(i,j)$ is equal to the number of ways to distribute $j-i$ apples to k people, we conclude that, if $1 \leq i \leq j \leq n$, then

$$M^k(i,j) = A(j-i,k) = \binom{j-i+k-1}{k-1}.$$

This solves our problem completely since the only nonzero entries of the matrix M^k are $M^k(i,j)$ with $1 \leq i \leq j \leq n$. □

Question 3. The 300×300 matrix A has all the main diagonal entries equal to 17 and the rest of the entries are 11. What is $\det(A)$?

Answer: Denote by J the 300×300 matrix whose entries are all equal to 1 and by I the 300×300 identity matrix. Then, the matrix A can be written as

$$A = 11J + 6I. \qquad (5.123)$$

Recall that the determinant of a matrix is equal to the product of all the eigenvalues of the matrix; see, for example, [6]. Thus, we will calculate the determinant of the matrix A by finding all of the eigenvalues of A and then taking their product.

5.4. LINEAR ALGEBRA

If μ and v are an eigenvalue of J and the corresponding eigenvector, then $Jv = \mu v$ and, from (5.123), we find that

$$Av = (11\mu + 6)v.$$

Thus,

$$\lambda = 11\mu + 6 \qquad (5.124)$$

is an eigenvalue of the matrix A with v being the corresponding eigenvector.

Moreover, all the eigenvalues and the eigenvectors of an $n \times n$ matrix with all entries equal to 1, including their multiplicities, are as follows:[2]

• The eigenvalue n has multiplicity 1. The corresponding eigenvector v_1 has all entries equal to 1;

• The eigenvalue 0 has multiplicity $n-1$. There are $n-1$ eigenvectors $v_2, \ldots, v_n$ that correspond to the eigenvalue 0. The vector v_i has the components $v_i(1) = 1$, $v_i(i) = -1$, and $v_i(j) = 0$ for all $j \notin \{1, i\}$.

Thus, the 300×300 matrix J with all entries equal to 1 has one eigenvalue equal to 300 and eigenvalue 0 with multiplicity 299. Then, from (5.124), it follows that the matrix A has one eigenvalue equal to 3306 and 299 eigenvalues equal to 6. We conclude that the determinant of the matrix A, which is equal to the product of all the eigenvalues of A, is

$$\det(A) = 3306 \cdot 6^{299}. \quad \square$$

Question 4. A bug starts at the vertex A of a triangle ABC. It then moves to one of its two adjacent vertices. How many paths of length 8 end back at vertex A? For example, one such path is $ABCABCABA$.

---
[2]This result is derived and used for Question 3 in Section 3.2 in the interview questions book [5].

Answer: The triangle on which the bug moves is a complete graph of size 3. Denote by $E = (e_{ij})_{1 \leq i,j \leq 3}$ the 3×3 incidence matrix of this graph. The entries of the matrix E are 0 or 1. The entry e_{ij} is 1 if there is an edge between the vertices i and j. Re–label the vertices A, B, and C of the triangle to 1, 2, and 3. The incidence matrix is

$$E = \begin{bmatrix} 0 & 1 & 1 \\ 1 & 0 & 1 \\ 1 & 1 & 0 \end{bmatrix}.$$

We will use the following result which was also used in Question 2:

If E is an incidence matrix of a graph, then $E^k(i,j)$ is the total number of paths of length k from the vertex i to the vertex j.

Thus, the number of paths of length 8 starting at A and ending back at A is equal to the entry $(1,1)$ of the matrix E^8.

To calculate E^8, note that, since the matrix E is symmetric, it has the diagonal form

$$E = QDQ^t \qquad (5.125)$$

where Q is the orthogonal matrix whose columns are the eigenvectors of the matrix E, and D is a diagonal matrix whose diagonal entries are the eigenvalues of E. Then,

$$E^8 = QD^8Q^t. \qquad (5.126)$$

We now compute the matrices D and Q. The characteristic polynomial of the matrix E is

$$\begin{aligned} P_E(t) &= \det(tI - E) = \det \begin{bmatrix} t & -1 & -1 \\ -1 & t & -1 \\ -1 & -1 & t \end{bmatrix} \\ &= t^3 - 2 - 3t = (t-2)(t+1)^2. \end{aligned}$$

5.4. LINEAR ALGEBRA

The roots of $P_E(t)$ are 2, with multiplicity 1 and -1, with multiplicity 2. Since the eigenvalues of a matrix are the roots of its characteristic polynomial, it follows that the eigenvalues of the matrix E are $\lambda_1 = 2$, $\lambda_2 = \lambda_3 = -1$. The corresponding eigenvectors are

$$v_1 = \begin{bmatrix} \frac{1}{\sqrt{3}} \\ \frac{1}{\sqrt{3}} \\ \frac{1}{\sqrt{3}} \end{bmatrix} ; \quad v_2 = \begin{bmatrix} \frac{1}{\sqrt{2}} \\ -\frac{1}{\sqrt{2}} \\ 0 \end{bmatrix} ; \quad v_3 = \begin{bmatrix} \frac{1}{\sqrt{6}} \\ \frac{1}{\sqrt{6}} \\ -\frac{2}{\sqrt{6}} \end{bmatrix}.$$

Thus, the matrices Q and D are

$$Q = \begin{bmatrix} \frac{1}{\sqrt{3}} & \frac{1}{\sqrt{2}} & \frac{1}{\sqrt{6}} \\ \frac{1}{\sqrt{3}} & -\frac{1}{\sqrt{2}} & \frac{1}{\sqrt{6}} \\ \frac{1}{\sqrt{3}} & 0 & -\frac{2}{\sqrt{6}} \end{bmatrix} ; \quad D = \begin{bmatrix} 2 & 0 & 0 \\ 0 & -1 & 0 \\ 0 & 0 & -1 \end{bmatrix}.$$

Then, from (5.126), we obtain that

$$E^8 = Q \begin{bmatrix} 2^8 & 0 & 0 \\ 0 & (-1)^8 & 0 \\ 0 & 0 & (-1)^8 \end{bmatrix} Q^t.$$

By direct computation, we find that $E^8(1,1) = 86$ and therefore conclude that there are 86 paths of length 8 that start at the vertex A and end back at A. $\square$

Question 5. In determinant tic–tac–toe, Player 1 enters a 1 in an empty 3×3 matrix. Player 0 counters with a 0 in a vacant position, and the play continues in turn until the 3×3 matrix is completed with five 1s and four 0s. Player 0 wins if the determinant is 0, otherwise Player 1 wins. Assuming that both players pursue optimal strategies, who will win and how?

Answer: Player 0 always wins, regardless of Player 1's strategy!

We will use the standard labeling $\{a_{i,j}\}$, $i,j = 1, 2, 3$, for the entries of the 3×3 matrix to show the winning strategy for Player 0.

The determinant of a matrix only changes its sign if the rows (or columns) are permuted. Therefore, without loss of generality, we can assume that Player 1 sets $a_{1,1} = 1$ in the first move.

Will will show that Player 0 can win if he (or she) counters this first move of Player 1 by placing $a_{2,2} = 0$. There are only 4 cases to consider for Player 1's next move, up to symmetry:

• If Player 1 sets $a_{1,2} = 1$ in the next move, Player 0 counters with $a_{2,3} = 0$, which forces Player 1 to set $a_{2,1} = 1$, in order to protect against the second row having all 0's:[3]

$$\begin{bmatrix} 1 & 1 & \star \\ \star & 0 & \star \\ \star & \star & \star \end{bmatrix} \implies \begin{bmatrix} 1 & 1 & \star \\ \star & 0 & 0 \\ \star & \star & \star \end{bmatrix} \implies \begin{bmatrix} 1 & 1 & \star \\ 1 & 0 & 0 \\ \star & \star & \star \end{bmatrix}$$

Player 0 now chooses $a_{3,3} = 0$, which forces Player 1 to set $a_{1,3} = 1$ to protect against the third column having all 0's. Player 0 then chooses $a_{3,2} = 0$:

$$\begin{bmatrix} 1 & 1 & \star \\ 1 & 0 & 0 \\ \star & \star & 0 \end{bmatrix} \implies \begin{bmatrix} 1 & 1 & 1 \\ 1 & 0 & 0 \\ \star & \star & 0 \end{bmatrix} \implies \begin{bmatrix} 1 & 1 & 1 \\ 1 & 0 & 0 \\ \star & 0 & 0 \end{bmatrix}$$

After Player 1 sets $a_{3,1} = 1$, the resulting matrix has determinant 0:

$$\det \begin{bmatrix} 1 & 1 & 1 \\ 1 & 0 & 0 \\ 1 & 0 & 0 \end{bmatrix} = 0$$

and Player 0 wins!

---
[3]Throughout this solution, $\star$ denotes an entry not yet set by either player.

5.4. LINEAR ALGEBRA

• If Player 1 sets $a_{1,3} = 1$ in the next move, Player 0 counters with $a_{3,2} = 0$, which forces Player 1 to set $a_{1,2} = 1$, in order to protect against the second column having all 0's:

$$\begin{bmatrix} 1 & \star & 1 \\ \star & 0 & \star \\ \star & \star & \star \end{bmatrix} \implies \begin{bmatrix} 1 & \star & 1 \\ \star & 0 & \star \\ \star & 0 & \star \end{bmatrix} \implies \begin{bmatrix} 1 & 1 & 1 \\ \star & 0 & \star \\ \star & 0 & \star \end{bmatrix}$$

Player 0 now chooses $a_{2,3} = 0$, which forces Player 1 to set $a_{2,1} = 1$ to protect against the second row having all 0's. Player 0 then chooses $a_{3,3} = 0$:

$$\begin{bmatrix} 1 & 1 & 1 \\ \star & 0 & 0 \\ \star & 0 & \star \end{bmatrix} \implies \begin{bmatrix} 1 & 1 & 1 \\ 1 & 0 & 0 \\ \star & 0 & \star \end{bmatrix} \implies \begin{bmatrix} 1 & 1 & 1 \\ 1 & 0 & 0 \\ \star & 0 & 0 \end{bmatrix}$$

After Player 1 sets $a_{3,1} = 1$, the resulting matrix has determinant 0:

$$\det \begin{bmatrix} 1 & 1 & 1 \\ 1 & 0 & 0 \\ 1 & 0 & 0 \end{bmatrix} = 0$$

and Player 0 wins again.

• If Player 1 sets $a_{2,3} = 1$ in the next move, Player 0 counters with $a_{3,2} = 0$, which forces Player 1 to set $a_{1,2} = 1$, in order to protect against the second column having all 0's:

$$\begin{bmatrix} 1 & \star & \star \\ \star & 0 & 1 \\ \star & \star & \star \end{bmatrix} \implies \begin{bmatrix} 1 & \star & \star \\ \star & 0 & 1 \\ \star & 0 & \star \end{bmatrix} \implies \begin{bmatrix} 1 & 1 & \star \\ \star & 0 & 1 \\ \star & 0 & \star \end{bmatrix}$$

Player 0 now chooses $a_{3,1} = 0$, which forces Player 1 to set $a_{3,3} = 1$ to protect against the third row having all 0's. Player 0 then chooses $a_{2,1} = 0$:

$$\begin{bmatrix} 1 & 1 & \star \\ \star & 0 & 1 \\ 0 & 0 & \star \end{bmatrix} \implies \begin{bmatrix} 1 & 1 & \star \\ \star & 0 & 1 \\ 0 & 0 & 1 \end{bmatrix} \implies \begin{bmatrix} 1 & 1 & \star \\ 0 & 0 & 1 \\ 0 & 0 & 1 \end{bmatrix}$$

After Player 1 sets $a_{1,3} = 1$, the resulting matrix has determinant 0:

$$\det \begin{bmatrix} 1 & 1 & 1 \\ 0 & 0 & 1 \\ 0 & 0 & 1 \end{bmatrix} = 0$$

and Player 0 wins once again.

• If Player 1 sets $a_{3,3} = 1$ in the next move, Player 0 counters with $a_{2,3} = 0$, which forces Player 1 to set $a_{2,1} = 1$, in order to protect against the second row having all 0's:

$$\begin{bmatrix} 1 & \star & \star \\ \star & 0 & \star \\ \star & \star & 1 \end{bmatrix} \implies \begin{bmatrix} 1 & \star & \star \\ \star & 0 & 0 \\ \star & \star & 1 \end{bmatrix} \implies \begin{bmatrix} 1 & \star & \star \\ 1 & 0 & 0 \\ \star & \star & 1 \end{bmatrix}$$

Player 0 now chooses $a_{1,2} = 0$, which forces Player 1 to set $a_{3,2} = 1$ to protect against the second column having all 0's. Player 0 then chooses $a_{1,3} = 0$:

$$\begin{bmatrix} 1 & 0 & \star \\ 1 & 0 & 0 \\ \star & \star & 1 \end{bmatrix} \implies \begin{bmatrix} 1 & 0 & \star \\ 1 & 0 & 0 \\ \star & 1 & 1 \end{bmatrix} \implies \begin{bmatrix} 1 & 0 & 0 \\ 1 & 0 & 0 \\ \star & 1 & 1 \end{bmatrix}$$

After Player 1 sets $a_{3,1} = 1$, the resulting matrix has determinant 0:

$$\det \begin{bmatrix} 1 & 0 & 0 \\ 1 & 0 & 0 \\ 1 & 1 & 1 \end{bmatrix} = 0$$

and Player 0 wins.

Thus, Player 0 has a strategy to win regardless of what Player 1 does. □

Question 6. The city has a total of n citizens. The citizens are forming various clubs, and at some point the government decided to implement the following rules:

5.4. LINEAR ALGEBRA

1) Each club must have an odd number of members.

2) Every two clubs must have an even number of members in common.

Show that it is impossible to form more than n clubs.

Answer: Label the citizens by $1, 2, \ldots, n$, and label their clubs by C_1, C_2, ..., C_m. Define the $m \times n$ matrix $M = (m_{i,j})_{i=1:m, j=1:n}$ by $m_{i,j} = 1$ if citizen j is in club C_i, and $m_{i,j} = 0$ otherwise. Note that

$$\text{rank}(M) \leq n. \tag{5.127}$$

Consider the $m \times m$ matrix MM^t, whose entry in position (i, k) is $\sum_{j=1}^{n} m_{i,j} m_{k,j}$, which is exactly the number of citizens common to both clubs C_i and C_k. In what follows, all the operations will be done over the two-element field $\mathbb{F}_2 = \{0, 1\}$.

Since every two clubs have an even number of members in common, it follows that the (i, k) entry of MM^t is 0 (in $\mathbb{F}_2$), if $i \neq k$. Moreover, since every club has an odd number of members, the (i, i) entry of MM^t (which represents the number of members of the club C_i) is 1.

Then, $MM^t = I_m$, where I_m is the $m \times m$ identity matrix, and therefore $\text{rank}(MM^t) = m$. Recall that the rank of a product of two matrices is less than or equal to the minimum of the ranks of the two matrices, i.e., $\text{rank}(AB) \leq \min(\text{rank}(A), \text{rank}(B))$. Thus,

$$\begin{aligned} m &= \text{rank}(MM^t) \leq \min(\text{rank}(M), \text{rank}(M^t)) \\ &= \text{rank}(M), \end{aligned} \tag{5.128}$$

where we used the fact that the rank of the transpose of a matrix is the same as the rank of the matrix and therefore $\text{rank}(M^t) = \text{rank}(M)$.

From (5.127) and (5.128), we obtain that $m \leq n$ and conclude that the number of clubs (m) is at most the number of citizens (n). $\square$

Chapter 6

Appendix

6.1 Sums of the form $\sum_{k=1}^{n} k^i$

The following sums occur frequently in practice, e.g., when estimating the operation counts of numerical algorithms:

$$\sum_{k=1}^{n} k = \frac{n(n+1)}{2}; \tag{6.1}$$

$$\sum_{k=1}^{n} k^2 = \frac{n(n+1)(2n+1)}{6}; \tag{6.2}$$

$$\sum_{k=1}^{n} k^3 = \left(\frac{n(n+1)}{2}\right)^2, \tag{6.3}$$

where $n \geq 1$ is a positive integer.

Using mathematical induction, it is easy to show that formulas (6.1–6.3) are correct. For example, for formula (6.3), a proof by induction can be given as follows: if $n = 1$, both sides of (6.3) are equal to 1. We assume that (6.3) holds for a fixed positive integer n and prove that (6.3) also holds for $n + 1$. In other words, we assume that n is a fixed positive integer for which the following identity holds

$$\sum_{k=1}^{n} k^3 = \left(\frac{n(n+1)}{2}\right)^2. \tag{6.4}$$

We need to show that

$$\sum_{k=1}^{n+1} k^3 = \left(\frac{(n+1)(n+2)}{2}\right)^2. \tag{6.5}$$

From (6.4), and by a simple computation, we find that

$$\sum_{k=1}^{n+1} k^3$$
$$= \sum_{k=1}^{n} k^3 + (n+1)^3 = \left(\frac{n(n+1)}{2}\right)^2 + (n+1)^3$$
$$= (n+1)^2 \left(\frac{n^2}{4} + n + 1\right) = \frac{(n+1)^2(n^2 + 4n + 4)}{4}$$
$$= \left(\frac{(n+1)(n+2)}{2}\right)^2.$$

In other words, (6.5) is proven, and therefore (6.3) is established by induction for any positive integer $n \geq 1$.

While proving equalities (6.1–6.3) by induction is not difficult, an interesting question is how are these formulas derived in the first place? In other words, how do we find out what the correct right hand sides of (6.1–6.3) are?

One way to obtain closed formulas for any sum of the form

$$S(n, i) = \sum_{k=1}^{n} k^i, \tag{6.6}$$

where $i \geq 1$ is a positive integer, is to use the binomial formula

$$(a+b)^m = \sum_{j=0}^{m} \binom{m}{j} a^j b^{m-j}, \ \forall \, a, b \in \mathbb{R}, \tag{6.7}$$

6.1. SUMS OF POWERS

where m is a positive integer. The binomial coefficient $\begin{pmatrix} m \\ j \end{pmatrix}$ is given by

$$\begin{pmatrix} m \\ j \end{pmatrix} = \frac{m!}{j!\,(m-j)!}, \tag{6.8}$$

where the factorial of a positive integer k is defined as $k! = 1 \cdot 2 \cdots k$. Several elementary properties of the binomial coefficients are given below:

$$\begin{pmatrix} m \\ j \end{pmatrix} = \begin{pmatrix} m \\ m-j \end{pmatrix}, \quad \forall\, 0 \le j \le m;$$

$$\begin{pmatrix} m \\ 0 \end{pmatrix} = \begin{pmatrix} m \\ m \end{pmatrix} = 1, \quad \forall\, m \ge 1; \tag{6.9}$$

$$\begin{pmatrix} m \\ 1 \end{pmatrix} = \begin{pmatrix} m \\ m-1 \end{pmatrix} = m, \quad \forall\, m \ge 1. \tag{6.10}$$

Using the binomial formula (6.7) for $a = k$, $b = 1$, and $m = i+1$, where k and i are positive integers, we obtain

$$(k+1)^{i+1} = \sum_{j=0}^{i+1} \begin{pmatrix} i+1 \\ j \end{pmatrix} k^j$$

$$= k^{i+1} + \sum_{j=0}^{i} \begin{pmatrix} i+1 \\ j \end{pmatrix} k^j.$$

In the last equation we used that $\begin{pmatrix} i+1 \\ i+1 \end{pmatrix} = 1$. This was obtained from (6.9) by substituting $m = i+1$. Thus,

$$(k+1)^{i+1} - k^{i+1} = \sum_{j=0}^{i} \begin{pmatrix} i+1 \\ j \end{pmatrix} k^j. \tag{6.11}$$

Writing (6.11) for all positive integers $k = 1 : n$, summing over k, and using the notation from (6.6), we obtain that

$$(n+1)^{i+1} - 1$$
$$= \sum_{k=1}^{n} \left((k+1)^{i+1} - k^{i+1} \right)$$
$$= \sum_{k=1}^{n} \left(\sum_{j=0}^{i} \binom{i+1}{j} k^j \right)$$
$$= \sum_{j=0}^{i} \left(\sum_{k=1}^{n} \binom{i+1}{j} k^j \right)$$
$$= \sum_{j=0}^{i} \binom{i+1}{j} \left(\sum_{k=1}^{n} k^j \right)$$
$$= \sum_{j=0}^{i} \binom{i+1}{j} S(n,j)$$
$$= (i+1)S(n,i) + \sum_{j=0}^{i-1} \binom{i+1}{j} S(n,j); \quad (6.12)$$

for the last equality, we used the fact that $\binom{i+1}{i} = i+1$; cf. (6.10).

Solving for $S(n,i)$ in (6.12), we obtain the following recursive formula for $S(n,i) = \sum_{k=1}^{n} k^i$:

$$S(n,i) = \frac{1}{i+1} \left((n+1)^{i+1} - 1 \right. \quad (6.13)$$
$$\left. - \sum_{j=0}^{i-1} \binom{i+1}{j} S(n,j) \right),$$

for all $i \geq 1$. Since, for $i = 0$, it is easy to see that $S(n,0) = n$, the recursive formula (6.13) can be used to compute closed formulas for $S(n,i)$, for any positive integer $i \geq 1$.

6.1. SUMS OF POWERS

For example, to compute $S(n,1) = \sum_{k=1}^{n} k$, let $i = 1$ in formula (6.13). Since $S(n,0) = n$, we obtain that

$$\begin{aligned} S(n,1) &= \frac{1}{2}\left((n+1)^2 - 1 - \sum_{j=0}^{0}\binom{2}{j}S(n,j)\right) \\ &= \frac{1}{2}\left((n+1)^2 - 1 - \binom{2}{0}S(n,0)\right) \\ &= \frac{1}{2}\left((n+1)^2 - 1 - n\right) = \frac{n(n+1)}{2}, \end{aligned}$$

which is the same as formula (6.1).

To compute $S(n,2) = \sum_{k=1}^{n} k^2$, let $i = 2$ in formula (6.13) and use the facts that $S(n,0) = n$ and $S(n,1) = \frac{n(n+1)}{2}$ to obtain that

$$\begin{aligned} S(n,2) &= \frac{1}{3}\left((n+1)^3 - 1 - \sum_{j=0}^{1}\binom{3}{j}S(n,j)\right) \\ &= \frac{1}{3}\left((n+1)^3 - 1 - S(n,0) - 3S(n,1)\right) \\ &= \frac{1}{3}\left((n+1)^3 - 1 - n - \frac{3n(n+1)}{2}\right) \\ &= \frac{1}{3} \cdot \frac{2n^3 + 3n^2 + n}{2} = \frac{n(n+1)(2n+1)}{6}, \end{aligned}$$

which is the same as formula (6.2).

Bibliography

[1] B. Bollobas *The Art of Mathematics: Coffee Time in Memphis*, Cambridge University Press, 2006.

[2] S. Boyd and L. Vandenberghe *Convex Optimization*, Cambridge University Press, 2004.

[3] J. Justicz, E. Scheinerman, and P. Winkler *Random Intervals*, Amer. Math. Monthly, 97 (10): 881–889.

[4] I. Matić, R. Radoičić, and D. Stefanica *Probability and Stochastic Calculus Quant Interview Questions*. FE Press, 2021.

[5] D. Stefanica, R. Radoičić, and T.-H. Wang *150 Most Frequently Asked Questions on Quant Interviews*. Second edition, FE Press, 2019.

[6] D. Stefanica. *A Linear Algebra Primer for Financial Engineering*. FE Press, 2014.

[7] D. Stefanica. *A Primer for the Mathematics of Financial Engineering.* Second edition, FE Press, 2011.

[8] D. West. *Introduction to Graph Theory*. Second edition, Pearson, 2017.

Made in the USA
Middletown, DE
07 October 2024